Mechanisms of cooperativity and allosteric regulation in proteins

Mechanisms of cooperativity and allosteric regulation in proteins

MAX PERUTZ
Medical Research Council Laboratory of Molecular Biology, Cambridge

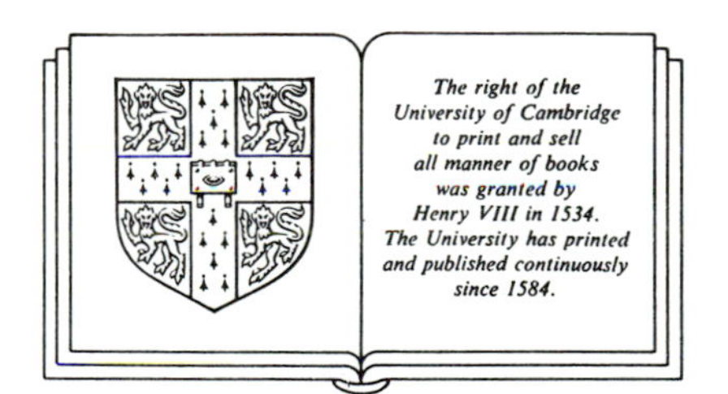

CAMBRIDGE UNIVERSITY PRESS
Cambridge
New York Port Chester
Melbourne Sydney

Published by the Press Syndicate of the University of Cambridge
The Pitt Building, Trumpington Street, Cambridge CB2 1RP
40 West 20th Street, New York NY 10011, USA
10 Stamford Road, Oakleigh, Melbourne 3166, Australia

First published 1989 in *Quarterly Reviews of Biophysics*
First published as a book 1990
Printed in Great Britain by the University Press, Cambridge

British Library cataloguing in publication data

Perutz, M. F.
Mechanisms of cooperativity and allosteric regulation in
proteins.
1. Organisms. Proteins. Biochemistry
I. Title II. Quarterly review of biophysics
574.19'245

Library of Congress cataloguing in publication data

Perutz, Max F.
Mechanisms of cooperativity and allosteric regulation in proteins
Max Perutz.
p. cm.
Includes bibliographical references.
ISBN 0 521 38648 9
1. Allosteric proteins. 2. Allosteric regulation. 3. Allosteric
enzymes. I. Title.
[DNLM: 1. Allosteric Regulation. 2. Hemoglobins–analysis.
3. Proteins–metabolism. QU 55 P471m]
QP552.A436P47 1990
574.19'245–dc20 89-25471 CIP

ISBN 0 521 38648 9

UP

To the memory of Jacques Monod, and to François Jacob, Jeffries Wyman and Daniel Koshland whose imaginative theories and elegant experiments gave the initial impetus for much of the work described here.

Contents

Preface

In the popular view the structure of DNA has told us all about the molecular basis of life, but in fact DNA's and most RNA's are chemically inert, whereas proteins are the workhorses of the living cell. They function as catalysts and genetic regulators, pumps and motors, receptors and transducers, stores and transporters, scaffolds and walls, toxins and antitoxins, conductors and insulators, and much else. They achieve their fantastic versatility with a repertoire of only twenty amino acids, the same in all organisms from archebacteria to mammals.

Many proteins are quiescent unless activated by specific chemical stimuli. The first hint of such stimuli being capable of triggering a change in structure of a protein came in 1938 from an observation by Felix Haurowitz, then a young professor of biochemistry at the Charles University in Prague. He crystallized oxyhaemoglobin of horse in salt-free water, in which it is sparingly soluble, and obtained a mush of scarlet needles. He put it away in a refrigerator and forgot about it. When he found it again a few weeks later, it had turned into a purple soup, because bacteria had used up all the oxygen and reduced the haemoglobin. At the bottom of the flask Haurowitz saw purple, hexagonal plates of deoxyhaemoglobin. He put a suspension of these crystals on a slide, placed a cover slip over it and watched it under the microscope.

As oxygen penetrated the solution, its colour gradually changed from purple back to scarlet, starting from the edge of the liquid and proceeding inwards. At the same time the purple crystals dissolved and scarlet needles grew in their place. The figure shows Haurowitz's original photograph. He concluded that the crystallization of deoxy- and oxyhaemoglobin in different forms implied that haemoglobin must change its structure every time it takes up or releases oxygen. He sent an account of his observation to Hoppe-Seyler's *Zeitschrift für physiologische Chemie*, the journal where all his previous work had appeared, and where it was published a few months later (Haurowitz, 1938).

Nearly fifty years later I was in the United States and telephoned Haurowitz at his home in Bloomington, Indiana, to ask if he would like me to visit him. He told me to come, because he had something very important to tell me. He was then over ninety and seriously ill, which made me wonder if he had some last message to give me about his theories of immunology. In fact, he talked about that paper. He told me that when he sent it to Hoppe-Seyler's *Zeitschrift*, German journals were forbidden to publish papers by Jewish authors like himself, but the editors, F. Knoop in Freiburg and K. Thomas in Leipzig, ignored that prohibition and published his paper regardless of the Nazis' racial laws. Haurowitz said that Knoop was actually imprisoned for this offence.

Haurowitz died a few months after my visit. This preface gives me the opportunity of fulfilling his last wish that I should pay tribute to Knoop and Thomas's courageous stand in the face of the Nazi terror, which secured the publication of the first experimental evidence of an allosteric change in a protein.

The paper received little attention at first, but I noted it because Haurowitz was married to a cousin of mine and had actually started me off on my life-long X-ray

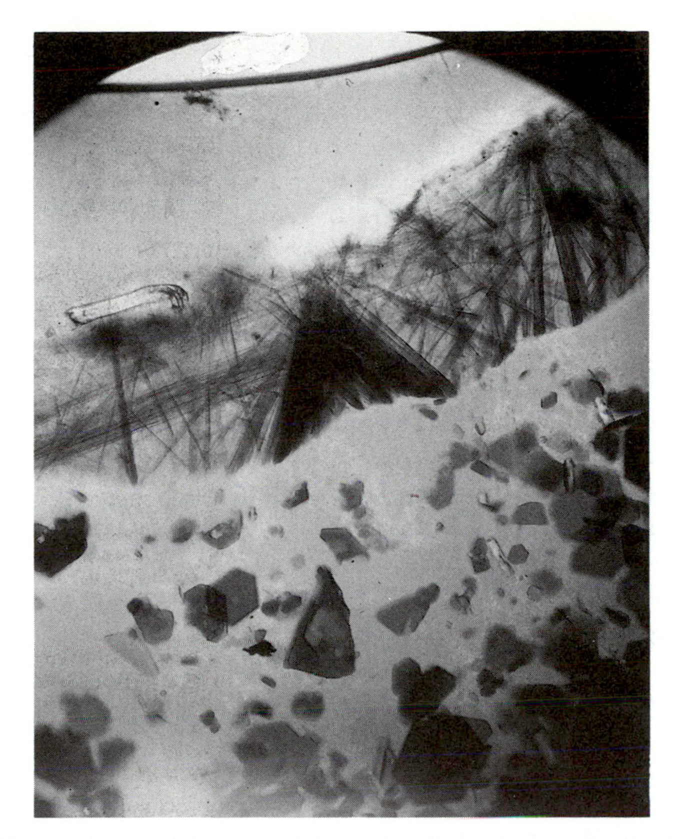

Suspension of human haemoglobin crystals between slide and cover slip. The lower part shows the purple hexagonal plates of deoxyhaemoglobin. The top of the photograph shows the meniscus between solution and air, below which needles of oxyhaemoglobin are seen to grow downwards into the deoxyhaemoglobin solution (Haurowitz, 1938).

work on haemoglobin the previous year. I repeated his experiment, but was disappointed that the crystals had the unit cell dimensions $a = 56$ Å, $c = 354$ Å, far too large for the X-ray techniques of those early days, and I therefore returned to my favourite crystals of horse oxy- and methaemoglobin. In the mid-1950s, Martin and Margaret Jope, biochemists then at Oxford, asked me to try X-ray work on crystals of human deoxyhaemoglobin precipitated, not from water, but from ammonium sulphate solutions. One of their polymorphic forms belonged to the space group $P2_1$ with two molecules in the unit cell, ideal for X-ray analysis. When a young physics graduate, Hilary Muirhead, joined me as a research student, I suggested that she might take up the structure of these crystals for her Ph.D. In October 1962 she obtained the first electron density map at 5.5 Å resolution, and we quickly built a model of it. Comparison with the model of horse oxy- or methaemoglobin built three years earlier showed that the subunits had moved apart on deoxygenation. The haemoglobin molecule had changed its quaternary structure (Muirhead & Perutz, 1963). Monod and I had been in touch all that time; he had already been excited a year earlier on hearing that a difference Patterson projection had shown the mercury atoms attached to the reactive sulphydryl groups to move apart by 6 Å on deoxygenation. That observation had given the first hint that a change of quaternary structure was taking place, and Monod was delighted to see it confirmed.

At that time my thoughts were concentrated on the X-ray analysis of haemoglobin. I cannot remember having read either Yates & Pardee's (1956) paper on feedback inhibition in aspartate transcarbamoylase, or Umbarger & Brown's (1958) and Changeux's (1961) papers on threonine deaminase. If I did, the analogy between the inhibition of these enzymes by the end-products of their biosynthetic pathways and the inhibition of haemoglobin by hydrogen ions did not occur to me. That was Monod's crucial idea, the seed that ripened into the two great papers on allostery (Monod *et al.* 1963; Monod *et al.* 1965).

Lysozyme was the first enzyme structure to be solved (Blake *et al.* 1965). X-ray work on allosteric enzymes did not begin until several years later, and at first only either the R- or the T-structure of each enzyme was solved, so that their mechanisms remained obscure. 1988 became the *annus mirabilis* when suddenly both forms of three different allosteric enzymes and of the *trp* repressor saw the light of day. The emergence of these structures has provided the impetus for writing this book.

I hope that this spring heralds a long summer, because X-ray analysis of the allosteric proteins responsible for the most basic chemical processes of life has hardly begun. For example, the structures of NADH-Q reductase, cytochrome c reductase and cytochrome c oxidase which produce the proton gradient in the mitochondrial membrane; of the mitochondrial ATPase that transduces the electrochemical energy stored there into energy-rich phosphate bonds; and of the myosin ATPase which uses that energy for locomotion, are all unknown. These proteins are hard to crystallize, but the crystallization of the membrane-bound photosynthetic reaction centre from a purple bacterium encourages me to think that it will be done, and that one day we shall learn how chemical energy is generated and turned into motion, how neurotransmitters activate their receptors or how absorption of a single photon by one molecule of rhodopsin can block the flow of more than a million sodium ions through the plasma membrane of a visual receptor.

Cambridge, 1989

REFERENCES

BLAKE, C. C. F., KOENIG, D. F., MAIR, G. A., NORTH, A. C. T., PHILLIPS, D. C. & SARMA, V. R. (1965) Structure of hen egg-white lysozyme. *Nature* **206**, 757–761.

HAUROWITZ, F. (1938) Das Gleichgewicht zwischen Hämoglobin und Sauerstoff. *Hoppe-Seyler Z. physiol. Chem.* **254**, 266–274.

MUIRHEAD, H. & PERUTZ, M. F. (1963) Structure of haemoglobin. A three-dimensional Fourier synthesis of reduced human haemoglobin at 5.5 Å resolution. *Nature* **199**, 633–639.

1. INTRODUCTION

Allosteric proteins control and coordinate chemical events in the living cell. When Monod conceived that idea he said that he had discovered the second secret of life. The first was the structure of DNA. The theory as published by Monod *et al.* (1963) was concerned chiefly with cooperativity and feedback inhibition of enzymes, such as the inhibition of threonine deaminase, the first enzyme in the pathway of the synthesis of isoleucine, by isoleucine, and its activation by valine. Two years later the theory was formalized by Monod *et al.* (1965).

It says that cooperative substrate binding, and modification of enzymic activity by metabolites bearing no stereochemical relationship to either substrate or product, may arise in proteins with two or more structures in equilibrium. It predicts that such proteins are likely to be made up of several subunits symmetrically arranged, and that the structures would differ by the arrangement of the subunits and number and/or energy of the bonds between them. In one structure the subunits would be constrained by strong bonds that would resist the tertiary change needed for substrate binding; they called this T for tense. In the other structure these constraints would be relaxed, and they called it R. In the transition between them, the symmetry of the molecule would be conserved, so that the activity of all its subunits would be either equally low or equally high. This assumption allowed the formulation of a mathematical model with only three independent variables: K_T and K_R, the association constants of the ligand with the protein in the T and R structures, and $L_0 = [T]/[R]$, the concentration ratio of the two structures in the absence of ligand. The postulate of symmetry also seemed to have some aesthetic fascination for Monod, like Ptolemy's spheres, quite independent of its formal advantages.

Koshland *et al.*'s (1966) sequential model did not invoke any restrictions. Each subunit is allowed to change its tertiary structure on substrate binding and thereby to affect the chemical activities of its neighbours.

According to Monod *et al.* (1963), allosteric enzymes would have the biological advantage that no direct interaction need occur between the substrate of the protein and the regulatory metabolite which controls its activity, because control would be due entirely to a change of structure induced in the protein when it binds its specific effector. They added the prophetic sentence: 'More complete observations, once available, might justify the conclusion that allosteric transitions frequently involve alterations in quaternary structure.'

The changes in quaternary structure that conserve the symmetry of oligomers can be defined by a theorem according to which any motion of a body can be described by a rotation about and translation along a suitably orientated axis. The two subunits of a dimer are related by a twofold axis of symmetry that brings them into congruence by a rotation of 180°. If that symmetrical equivalence of the subunits is to be preserved, the axis defining the motion of one of the monomers must be related to the axis defining the motion of the other monomer by the same twofold symmetry axis (Fig. 1). Glycogen phosphorylase is a dimeric enzyme that undergoes an allosteric transition of that kind. Similar rules hold for assemblies of 3, 4, 5 or 6 or more subunits, arranged in rings about a central rotation axis of symmetry. If equivalence is to be preserved, the axes defining the motion of the individual subunits must be related by the central rotation axes. Fig. 2 shows the case of an allosteric trimer.

If one dimer is inverted and placed over another, the four subunits are related by three mutually perpendicular axes of twofold symmetry (Fig. 3). If the monomers are rigidly linked in pairs, then the only allowed change in quaternary structure that preserves symmetry is a rotation about and translation along one of the twofold axes of one dimer relative to the other dimer. Vertebrate haemoglobins and bacterial phosphofructokinase function according to this model. If the subunits are not linked in pairs, each subunit can rotate about and shift along a different axis, but the four axes must be placed so that they are brought into congruence by the three twofold symmetry axes of the tetramer.

Certain enzymes are made up of one ring of identical subunits that is inverted and placed on top of another ring. Such an arrangement offers a ready way of allosteric control if the active sites are located at subunit boundaries. Any effector that turns one ring relative to the other ring could also turn each subunit about its own axis and thus shift the two faces of the active site relative to each other (Fig. 4). Glutamine synthetase may function like that, but so far no allosteric transition has been observed in that protein. More complex rotations and translations of subunits are possible in octahedral and icosahedral structures. Tomato Bushy Stunt Virus undergoes a concerted transition in which the icosahedral symmetry of its 180 subunits is preserved.

Cooperativity and feedback control can also arise by linkage between reactivity and reversible dissociation of oligomers. Suppose the affinity for a reactant is low in the oligomer and high in the monomer. If combination of any one monomer with reactant loosens the bonds between all the subunits in the oligomer, it would bias the equilibrium towards the dissociated form, whence the affinity would rise with rising concentration of reactant. Lamprey haemoglobin functions like that.

In allosteric proteins cooperativity arises not primarily by any direct interaction between the active sites, but mainly by a change in equilibrium between alternative structures at successive binding steps of the reactants or effectors. Fig. 5 shows sigmoid oxygen equilibrium curves of haemoglobin solutions with

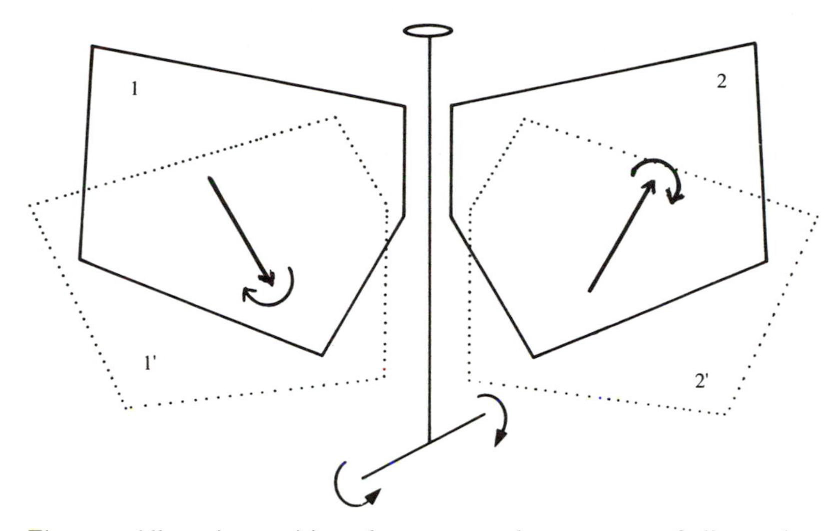

Fig. 1–4. Allosteric transitions that preserve the symmetry of oligomeric proteins.
Fig. 1. Possible changes of quaternary structure in a dimer of two identical subunits related by an axis of twofold symmetry, also called a dyad (symbol on top) (Point group C_2 or 2). In the simplest transition the left subunit turns anticlockwise about an axis normal to the dyad and pointing towards the observer; the right subunit turns anticlockwise about a colinear axis pointing away from the observer. More generally, the subunits can turn about any pair of axes related by the molecular dyad, for example the ones shown as bold arrows. They point into the picture, away from the observer.

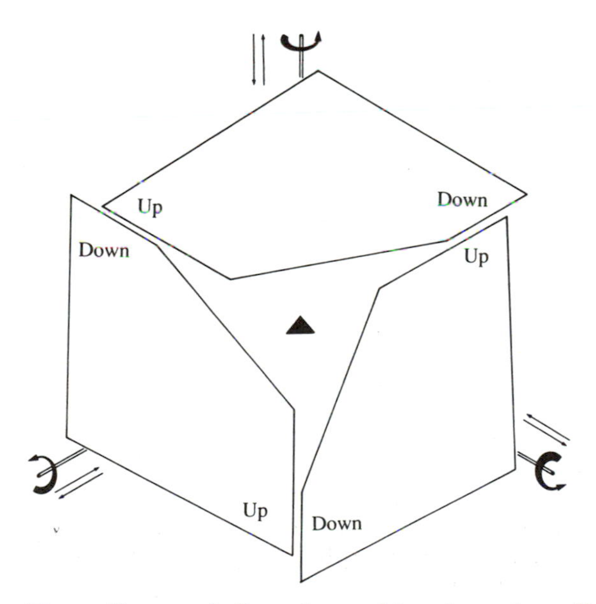

Fig. 2. Concerted allosteric transitions in a trimer (Point group C_3 or 3). In the simplest transition, each subunit can turn about and shift along an axis normal to the triad, indicated by the black triangle. In general, they can turn about and shift along any three axes related by the central triad axis of symmetry. Similar rules apply to rings of higher symmetry.

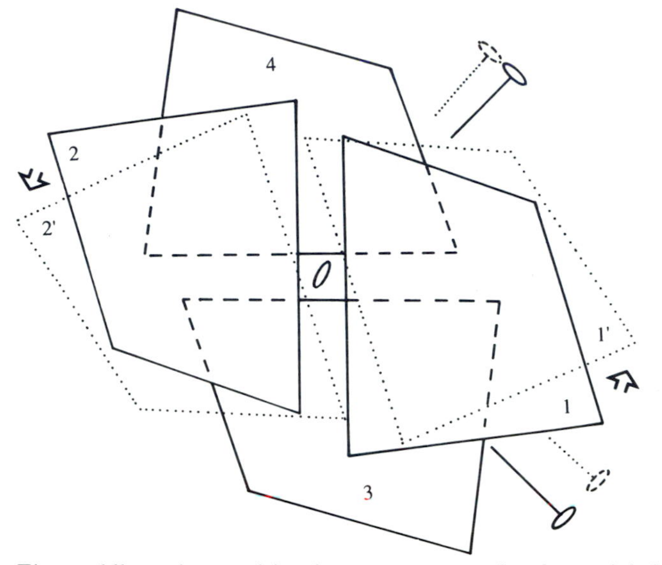

Fig. 3. Allosteric transition in a tetramer made of two tightly linked dimers facing each other. The tetramer has orthorhombic (D_2 or 222) symmetry; its subunits are related by three mutually perpendicular dyads. One dimer can turn relative to the other dimer about their common dyad; it can also shift along the dyad. Compare Figs. 9, p. 11, and 32, p. 45.

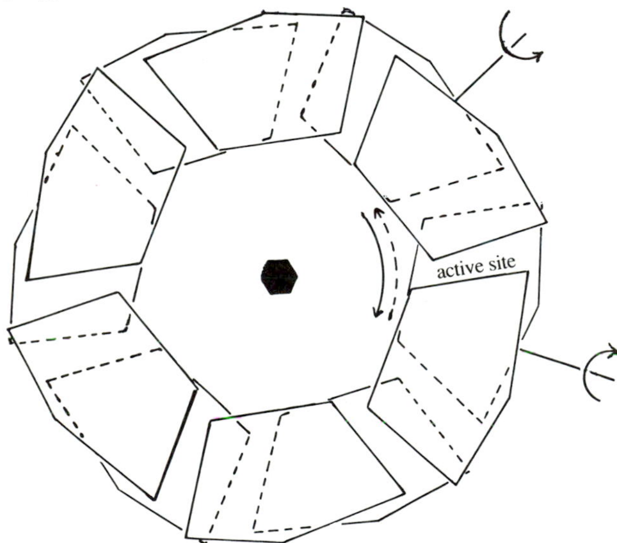

Fig. 4. Concerted allosteric transition in a dodecamer made of two rings of six identical subunits facing each other (point group symmetry D_6 or 62). One ring can turn relative to the other about the central hexad and it can also shift along it. In addition, each pair of subunits can turn about and shift along their common dyad. The same rules would apply to other even-numbered rings. In theory, the subunits could turn about and shift along any set of axes related by the point group symmetry of the oligomer. In practice, any set of movements other than those indicated here are likely to make them clash. Compare Figs. 39, p. 59, and 49 (Plate 6). In aspartate transcarbamylase (Fig. 39) one catalytic trimer turns about and shifts along the central triad relative to the other catalytic trimer, while the regulatory dimers turn about and shift along the dyads normal to the triad.

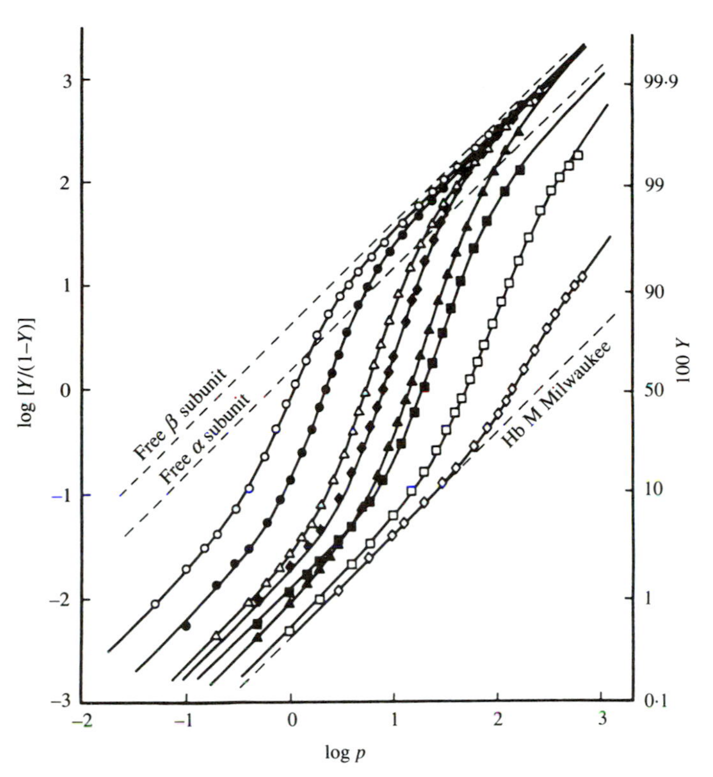

Fig. 5. Oxygen equilibrium curves of haemoglobin solutions containing different concentrations of allosteric effectors: H^+, Cl^-, CO_2, 2,3-diphosphoglycerate and inositolhexaphosphate, (From Imai, 1982.) Haemoglobin M Milwaukee remains in the T-structure even when saturated with oxygen.

different concentrations of allosteric effectors. All the curves are sigmoid, but the degree of cooperativity varies. At high pH in the absence of effectors the curve approaches that of non-cooperative free subunits. At low pH with strong effectors it approaches that of a non-cooperative mutant haemoglobin with low oxygen affinity. The degree of cooperativity, expressed as the slope at the midpoint of the sigmoid binding curves, is known as Hill's coefficient n. It is maximal if the ratio of [T]/[R] is unity when half the active sites have bound reactant. A plot of n against the allosteric constant L_0 follows a bell-shaped curve, n becoming unity when the protein remains in either the R or T structure regardless of the number of reactants bound (Fig. 19 of Baldwin, 1975 and Fig. 31, p. 44). Cooperativity ensures that most of the molecules in the solution are either fully oxygenated or fully deoxygenated. Fig. 6(*a*) shows the concentration of these two states and of intermediates in the T or R structure as a function of fractional saturation with ligands, calculated from allosteric theory (Baldwin, 1975). Figure 6(*b*) shows that the experimentally determined concentrations of haemoglobin intermediates are consistent with the predicted ones (Perrella *et al.* 1988).

Wyman realized that allosteric proteins are more complex than allowed for by Monod *et al.* (1965), because the chemical activities of each of several ligands may influence the affinities of the protein for all the others, so that all the equilibria are

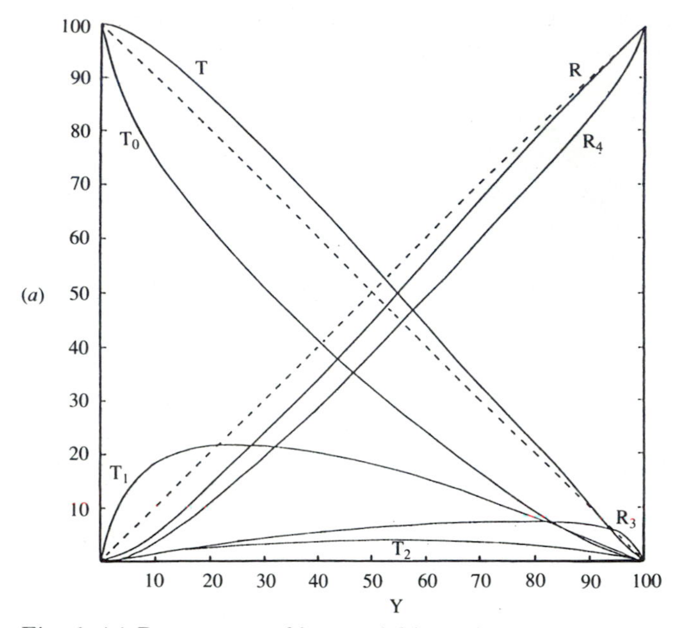

Fig. 6. (*a*) Percentages of haemoglobin molecules in the T and R structures as a function of fractional saturation Y with ligand, calculated from allosteric theory. T and R give the total percentages; T_0, T_1 and T_2 the percentages of Hb, Hb(O_2) and Hb$(O_2)_2$ in the T structure; R_3 and R_4 the percentages of Hb$(O_2)_3$ and Hb$(O_2)_4$ in the R-structure. The broken lines are straight. The curves are calculated for $L = 31\,600$ and $c = K_R/K_T = 0{\cdot}0562$, where K are association constants. (From Baldwin 1975.)

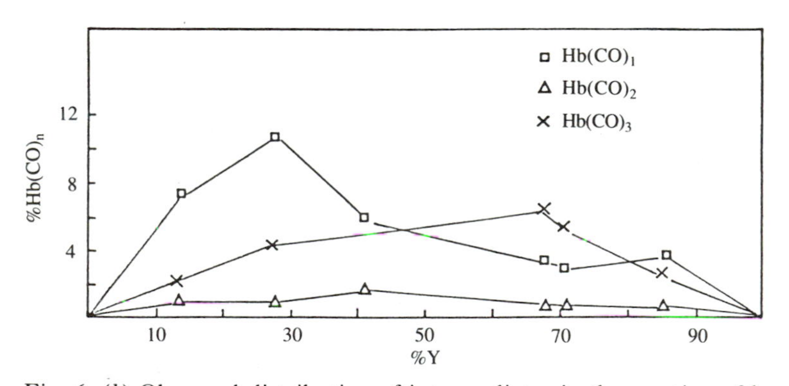

Fig. 6. (*b*) Observed distribution of intermediates in the reaction of haemoglobin with CO as a function of percentage saturation with CO. Note the similarity to the predicted one in Fig. 6(*a*). The reaction showed a Hill's coefficient of 3·0. The concentration of intermediates was obtained after quenching the reaction of two solutions expelled into a mixing chamber from two separate syringes. One syringe was filled with deoxy- and the other with carbonylhaemoglobin; alternatively one was filled with partially oxidized carbonylhaemoglobin and the other with a solution of dithionite. After mixing, the solution was expelled into a cryochamber at −25 °C containing a 10-fold molar excess of ferricyanide in an equal mixture of phosphate buffer and ethyleneglycol. Salt and ferricyanide were removed, and the different species separated by isoelectric focusing at −25 °C. Contrary to the claims of Gill *et al.* (1988), the triligated species is significantly populated. The diligated species are nearly all of the type $\alpha(CO)\beta(CO)\alpha\beta$. (From Perrella *et al.* 1986, 1988.)

linked. He developed a general theory based on a binding potential, L (for linkage), which is a function of the chemical potentials μ_i of all the components i, except the reference component (e.g. the protein itself). The potential L is defined by $\delta L / \delta \mu_i = n_i$, where n_i is the amount of component i per unit reference component. He derived equations that are related to the grand partition functions of statistical mechanics and express L in terms of the equilibria between two or more forms of the allosteric protein and their equilibria with ligands. Wyman's theory showed that the fineness of allosteric control increases with the number of binding sites for its effectors and that the shape of the ligand equilibrium curves varies with the chemical activities of the effectors (Wyman, 1967, 1984). Wyman and others recently extended his theory to proteins that polymerize into allosteric assemblies of higher order, so that two or more hierarchies of allosteric equilibria 'nest' in each other (Robert *et al.* 1987). The ligand binding properties of octopus haemocyanin, an oxygen-carrying protein with 70 active sites, have been interpreted in such terms (Connelly *et al.* 1989).

For some years haemoglobin had remained the only allosteric protein whose stereochemical mechanism was understood in detail, because the structures of both the T and R forms had been solved by X-ray analysis at high resolution. In 1988 this same happy state of affairs was reached with several enzymes and one genetic repressor. Haemoglobin has been widely regarded as the prototype of an allosteric protein, but is this true? Can there be cooperativity and feedback inhibition without a change of quaternary structure? Are changes in reactivity always linked to changes of the bonds between the subunits? Does the inactive state always correspond to the tense and the active one the relaxed structure? We shall see that cooperativity can arise also by purely entropic effects without changes in quaternary structure or reversible dissociation into subunits.

2. HAEMOGLOBIN: DEPENDENCE OF ALLOSTERIC EQUILIBRIUM ON SPIN STATE AND COORDINATION OF THE HAEM IRON

2.1 *Common features*

All haemoglobins have similar structures. The globin chain has a characteristic fold that envelops the haem in a deep pocket with its hydrophobic edges inside and its propionates facing the solvent. The chain is made up of seven or eight α-helical segments and an equal number of non-helical ones placed at the corners between them and at the ends of the chain (Fig. 7). According to a notation introduced by Watson & Kendrew (1961) the helices are named A to H, starting from the amino end; the non-helical segments that lie between helices are named AB, BC, CD, and so on. The non-helical segments at the ends of the chain are called NA and HC. Residues within each segment are numbered from the amino end: A1, A2, CD1, CD2, and so on. Evolution has conserved this fold of the chain despite great divergence of the sequence: the only residues common to all haemoglobins are the proximal histidine F8 and the phenylalanine CD1, which wedges the haem into its pocket. Most, but not all globins also have a histidine on the distal (oxygen) side of the haem. Ionized residues are excluded from the interior of the globin chains,

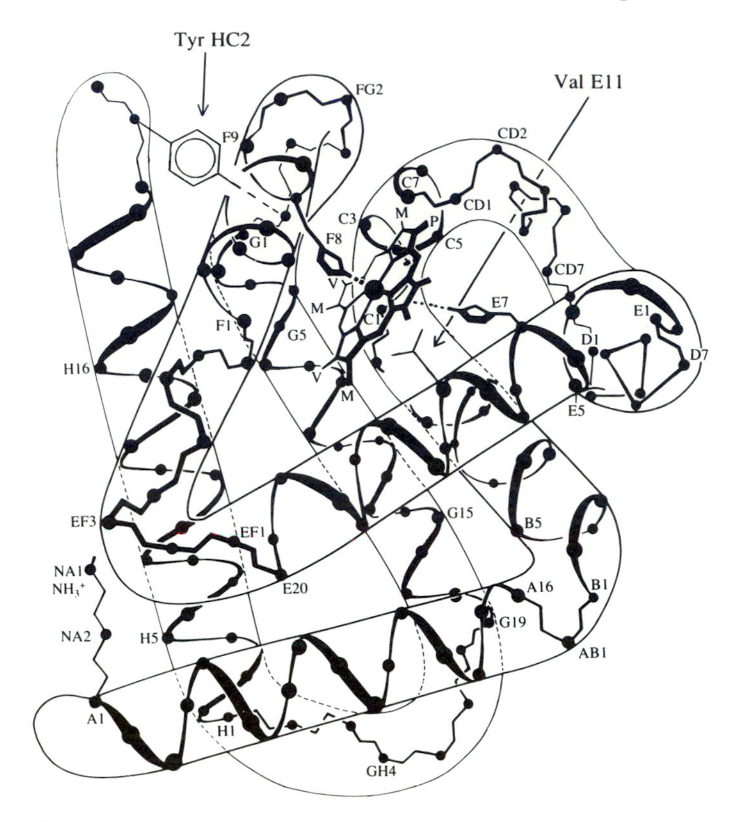

Fig. 7. Tertiary structure of β-chains of human haemoglobin, typical of haemoglobins and myoglobins of all other species. The figure also shows the proximal and distal histidines, marked F8 and E7, the distal valine E11 and the tryosine HC2 which ties down the C-terminus by its hydrogen bond to the main chain carbonyl of valine FG5.

which is filled largely by hydrocarbon side-chains, but some serines and threonines also occur there. The proximal and distal histidines (also called the haem-linked histidines) are potentially polar, but the proximal histidine does not ionize, and the pK_a of the distal one is so low ($\sim 5{\cdot}5$) that the fraction ionized *in vivo* is negligible.

2.2 *Lamprey haemoglobin: cooperativity by dissociation into subunits*

This is the simplest known allosteric protein. Its cooperative reaction with oxygen is based on reversible dissociation into subunits. In the absence of oxygen it forms dimers and tetramers with low oxygen affinity. Uptake of oxygen causes it to dissociate into monomers with high oxygen affinity. Thus the fraction of dimers and tetramers drops with rising oxygen saturation, and conversely, the oxygen affinity drops with rising haemoglobin concentration. There is also a marked Bohr effect, due to the cooperative uptake of protons on loss of oxygen (Briehl, 1963; Antonini & Brunori, 1971; Hendrickson & Love, 1971; Dohi *et al.* 1973).

The low oxygen affinity of dimeric or polymeric lamprey haemoglobin must be

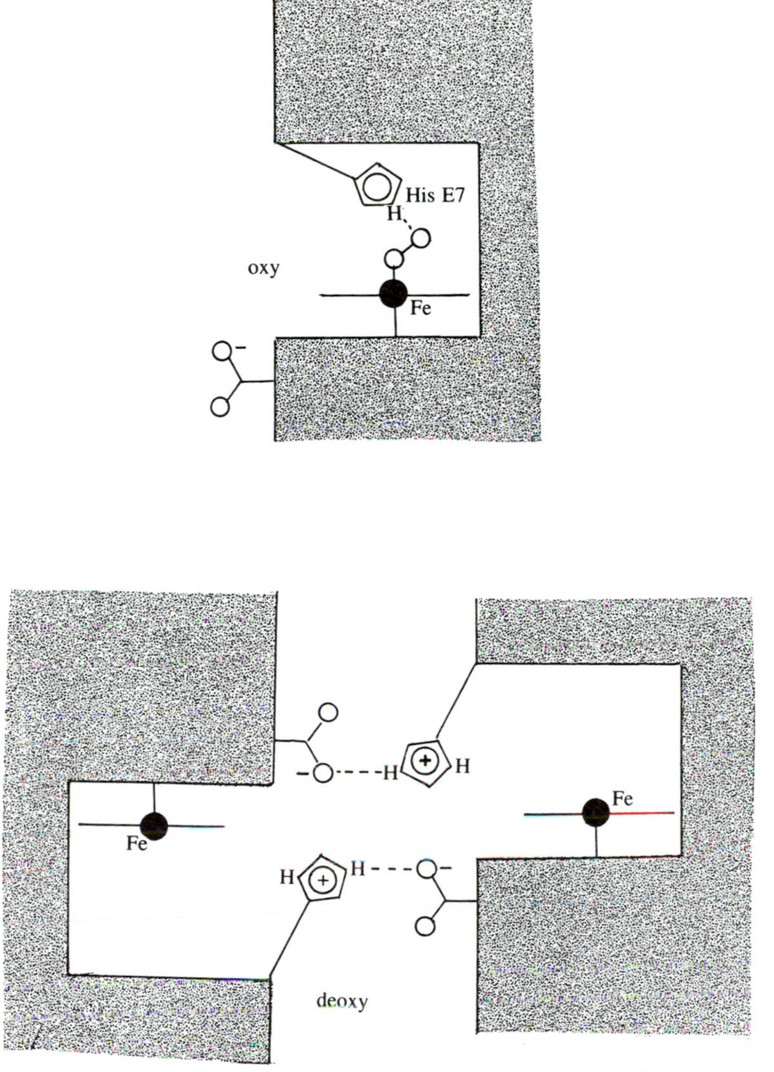

Fig. 8. Suggested mechanism of Bohr effect and cooperativity in lamprey haemoglobin. Oxyhaemoglobin is monomer; the distal histidine is in the haem pocket, hydrogen-bonded to the haem-linked oxygen. Deoxyhaemoglobin dimerizes; the distal histidine has swung out of the pocket and forms a hydrogen bond with a carboxylate of the neighbouring subunit, accompanied by the uptake of a proton.

due to constraints by bonds between its subunits. In the simplest case, reaction of one subunit with oxygen would break these bonds and dissociate a dimer into its two subunits, with the result that the oxygen affinity of the other subunit would be raised by the free energy equivalent of the monomer–dimer dissociation constant. We can infer the nature of at least one of these bonds from the Bohr effect. This appears to be due to ionization of groups with a pK_a of 6·5, which points to their being histidines, or less likely, α-amino groups. The N-terminal residue of lamprey haemoglobin is proline, and its amino group forms a hydrogen bond with a glutamate in the C-terminal helix (H8), so that it is not available, which leaves histidines as the only possible source. Lamprey haemoglobin has only two histidines: the proximal one which does not ionize, and the distal one. These facts suggest that the distal histidine alternates between two positions: an internal one in the haem pocket, where it forms a hydrogen bond with the haem-linked oxygen, and an external one, where it forms a hydrogen bond with an aspartate or glutamate of a neighbouring molecule (Fig. 8). In its internal position the histidine would have a low pK_a somewhere between 5·5 and 6·0; hydrogen-bonding to a carboxylic group would raise it to about 8·0. This change in pK_a would account for the large Bohr effect. The free energy of cooperativity would be provided by the energy of these two hydrogen bonds plus that of the other contacts between the two subunits. My interpretation could be tested by directed mutagenesis, replacing the distal histidine by a glutamine. This ought to inhibit the Bohr effect if my idea is valid. One alternative mechanism would consist in the formation of a hydrogen-bonded pair of carboxylates, one with a low and the other with a high pK_a. Such pairs exist at subunit contacts in tobacco mosaic virus (Caspar, 1963; Namba *et al.* 1989). However, the free energy needed to raise the pK_a of a glutamate or aspartate side-chain to 6·5 is of the order of 4 kcal/mol, about twice as much as the hydrogen bond energy gained in forming the pair, and the deficiency would have to be made up by other bonds between the subunits.

2.3 *Human haemoglobin*

2.3.1 *Oxygen equilibrium and allosteric transition*

Human haemoglobin is a tetramer composed of two α-chains, each containing 141 amino acid residues, and two β-chains each containing 146 residues. Each chain contains one haem. It binds oxygen cooperatively with a free energy of cooperativity of 3·5 kcal mol^{-1}, and its oxygen affinity is modulated by [H^+], [Cl^-], [CO_2] and [2,3-diphosphoglycerate] (DPG), known collectively as the heterotropic ligands (Fig. 5). At extreme dilutions (< 3 μM haem) cooperative effects arise as a result of reversible dissociation of tetrameric deoxy to dimeric oxyhaemoglobin ($\alpha_2\beta_2 \rightleftharpoons 2\alpha\beta$), but at higher haemoglobin concentrations the cooperative effects are due to an equilibrium between two alternative quaternary structures of the tetramer, the deoxy or T and the oxy or R structure.

The O_2 affinity of the R structure is slightly larger than the average one of free α- and β-subunits; that of the T-structure is lower by the equivalent of the free energy of cooperativity. The O_2 equilibrium can be described by the O_2 association constants K_T and K_R, usually expressed in (mmHg)$^{-1}$, and by the equilibrium constant $L_0 = [T]/[R]$ in the absence of O_2. Imai has shown empirically that log $K_T/K_R = A - 0{\cdot}25 \log L_0$, where A is a constant, which leaves K_R and K_T as the only independent variables. K_T varies over a wide range as a function of [H^+], [Cl^-], [CO_2] and [DPG]; K_R varies as a function of [H^+] below pH 7, but is little affected by the other ligands (Baldwin, 1975; Fermi & Perutz, 1981; Imai, 1982; Dickerson & Geis, 1983; Bunn & Forget, 1985; Angel *et al.* 1988; Perutz, 1979, 1987).

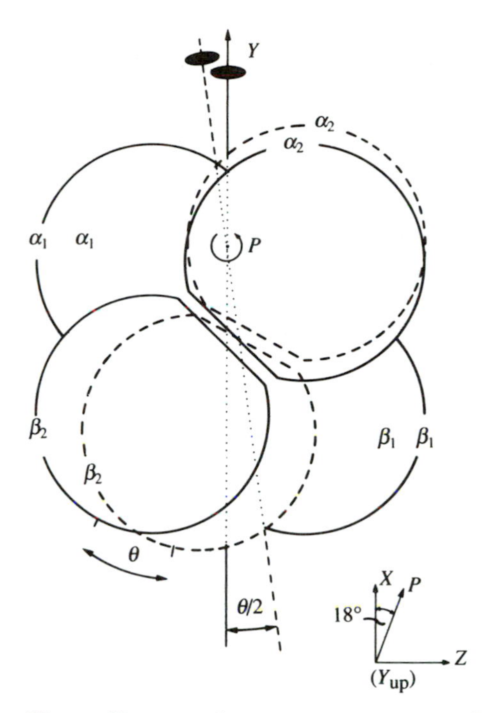

Fig. 9. Changes of quaternary structure of mammalian haemoglobin on transition from deoxy or T (full lines) to oxy or R (broken lines). The dimer $\alpha_2\beta_2$ turns by $\theta = 13°$ about the P axis; this entails a rotation of the dyad symmetry axis Y by $\theta/2 = 6{\cdot}5°$. If the four subunits were identical and the tetramer had 222 symmetry, P would have to coincide with X which points normal to the plane of the paper and towards the observer (from Baldwin & Chothia, 1979).

The T and R structures differ in the arrangement of the four subunits, referred to as the quaternary structure, and the conformation of the subunits, referred to as the tertiary structure. The quaternary $R \rightarrow T$ transition consists in a rotation of the dimer $\alpha_1\beta_1$ relative to the dimer $\alpha_2\beta_2$ by 12–15° and a translation of one dimer relative to the other by 0·8 Å. The $\alpha\beta$ dimers move relative to each other at the symmetry-related contacts $\alpha_1\beta_2$ and $\alpha_2\beta_1$ and at the contacts $\alpha_1\alpha_2$ and $\beta_1\beta_2$; the contacts $\alpha_1\beta_1$ and $\alpha_2\beta_2$ remain rigid (Fig. 9).

At the $\alpha_1\beta_2$ interface the non-helical segment $FG\alpha_1$ is in contact with helix $C\beta_2$ and helix $C\alpha_1$ with $FG\beta_2$. During the $R \rightarrow T$ transition, the contact $FG\alpha_1$-$C\beta_2$ acts as a ball-and-socket joint, while the contact $C\alpha_1$-$FG\beta_2$ acts as a two-way switch that shifts $C\alpha_1$ relative to $FG\beta_2$ by about 6 Å, like the knuckles of one hand moving over those of the other. Intermediate positions of the switch are blocked by steric hindrance. The gaps along the central cavity between α_1 and α_2 and between β_1 and β_2 narrow on transition from T to R. The shape of the $\alpha_1\beta_1$ and $\alpha_2\beta_2$ dimers is altered by changes in tertiary structure: for example, on oxygenation the distance between the α-carbons of residues $FG1\alpha_1$, and β_1 shrinks from 45·6 to 41·3 Å. These changes make an $\alpha_1\beta_1$ dimer that has the tertiary oxy structure a misfit in the quaternary T-structure, and an $\alpha_1\beta_1$ dimer that has the tertiary deoxy structure a misfit in the quaternary R-structure (see Figs. 10–15 of Baldwin & Chothia, 1979).

The key question for the understanding of haemoglobin function are these: how does the reaction with O_2 affect the stereochemistry at and around the haem so as to trigger the transition from the T to the R structure? What are the constraints of the T structure and how do they lower the O_2 affinity? By what mechanisms do the heterotropic ligands influence the O_2 affinity? Single-crystal X-ray analyses of deoxy- and oxyhaemoglobin, and of analogues of intermediates in the reactions with O_2 or CO, together with chemical, spectroscopic and magnetic studies, have furnished some of the answers.

2.3.2 *Stereochemical mechanism: changes on oxygenation in stereochemistry of the haems*

Fig. 10 summarizes the stereochemistry of the haems in deoxyhaemoglobins, oxyhaemoglobin and in two intermediates. In deoxyhaemoglobin the iron is high-spin ferrous (S = 2) and five-coordinated. The iron atoms are displaced from the planes of the porphyrin nitrogens, and the porphyrins are domed. On oxygenation the iron becomes low-spin ferrous (S = 0) and six-coordinated. The porphyrins flatten and the $Fe\text{-}N_{porphyrin}$ bond lengths contract from 2·06 to 1·98 Å, thus moving the iron atoms towards the porphyrin planes (Perutz, 1970, 1979). As a result the proximal histidines come 0·5–0·6 Å closer to the porphyrin planes in oxy- than in deoxyhaemoglobin. Do these movements trigger the allosteric transitions between the R and T structures, and if so, how are the transitions initiated?

Semi-liganded derivatives in the T-structure show that on combination of oxygen or carbonmonoxide with the α-haems the irons move by 0·15 Å towards the plane of the porphyrin nitrogens while the doming of the pyrroles is preserved. The movements of the irons are transmitted to the proximal histidines and their adjoining residues, while the bulk of the protein remains unperturbed. Thus perturbations are confined to what Gelin *et al.* (1983) have called the 'allosteric core'. On loss of iron-linked H_2O and reduction of the irons of a ferric haemoglobin in the R-structure, the iron atoms move away from the plane of the porphyrin nitrogens by 0·2 Å and the porphyrins become domed; the movements are transmitted not just to the proximal histidines and their adjoining residues, but also the the $\alpha_1\beta_2$ and $\alpha_2\beta_1$ contacts, which shift a short way towards their positions in the T-structure (Perutz *et al.* 1987; Liddington *et al.*1988).

There have been suggestions that the hydrogen bonds between N_δ of His F8 and the carbonyl of Leu F4 play a part in the allosteric mechanism. The lengths of these bonds may change in transition states, but they remain the same in deoxyhaemoglobin and oxyhaemoglobin. There has also been a suggestion that changes in charge transfer interactions between the porphyrin and Phe CD1 contribute to the free energy of cooperativity, but the distance between the phenylalanine side-chain and the porphyrin is too large (3·8–4·1 Å) for such interactions to occur. *We are thus left with the distances of the Fe's and the proximal*

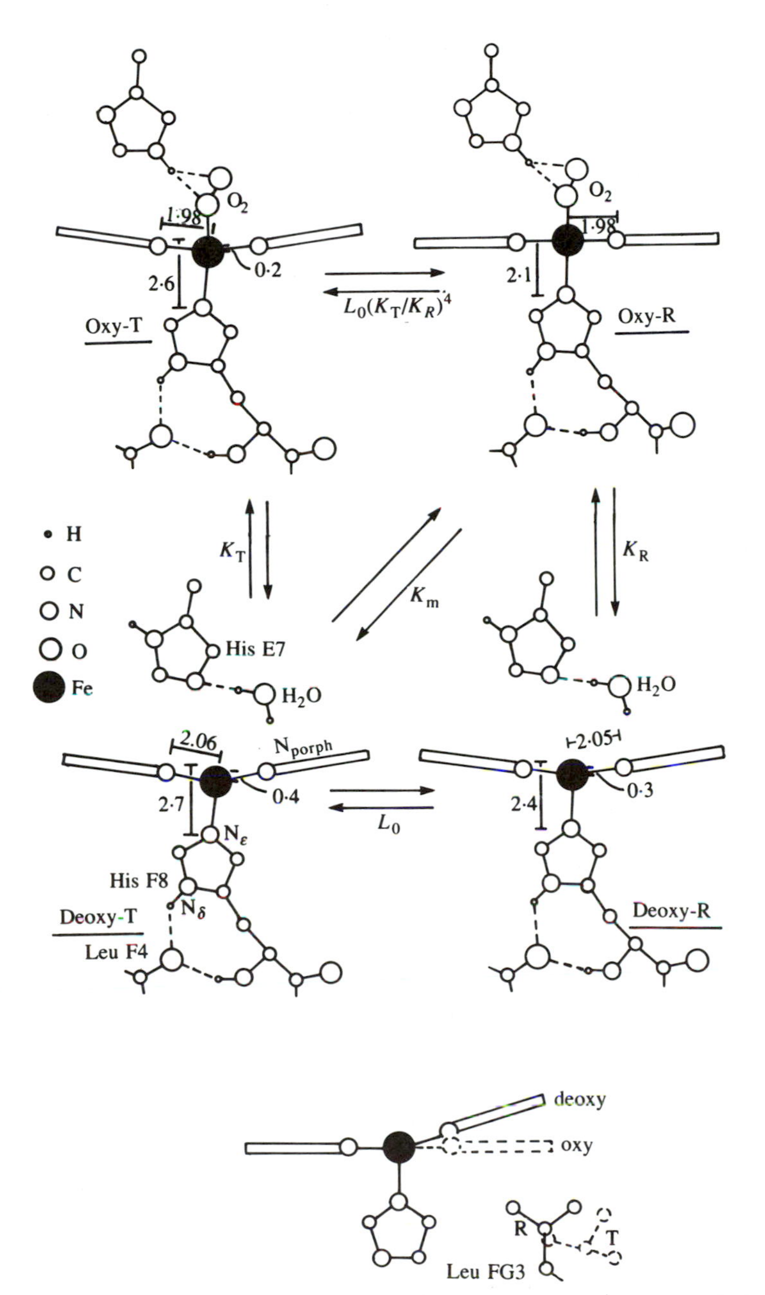

Fig. 10. Change of allosteric core on going from deoxy-T via oxy-T to oxy-R and from there via deoxy-R back to deoxy-T. The vertical bars indicate the distance of N_ϵ of the proximal histidine F8 from the mean plane of the porphyrin nitrogens and carbons (excluding β and γ carbons of the side-chains). The horizontal bar gives the $Fe\text{-}N_{porphyrin}$ distance, and the figure to the right of the iron atoms in the lower two diagrams the displacement of the iron from the plane of the porphyrin nitrogens. L_0 is the allosteric constant in the absence of oxygen, K_T and K_R are the association constants with oxygen of the T and R structures and K_m is the mean oxygen association constant. Note that the porphyrin is flat only in oxy-R and that the proximal histidine tilts relative to the haem normal in the T-structures. Note also the water molecule attached to the distal histidine in

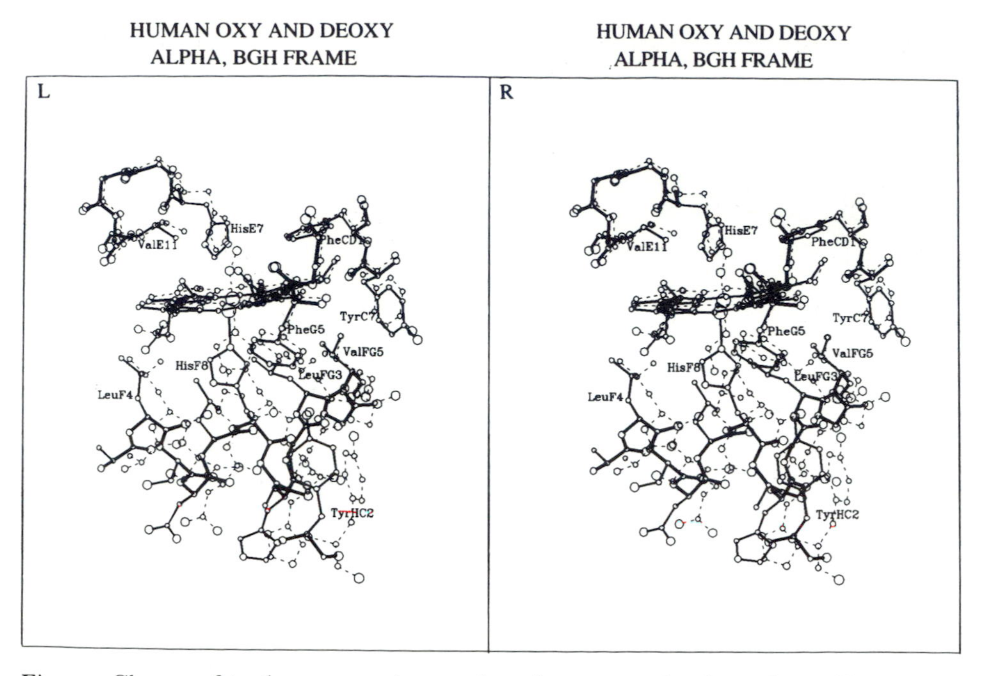

Fig. 11. Change of tertiary structure near the α-haem on going from deoxy-T (full lines) to oxy-R (broken lines). Note how the proximal histidine straightens and moves closer to the porphyrin, carrying the residues in helix F with it. (From Perutz *et al.* 1987.)

histidines from the porphyrin as the only determinants of the allosteric equilibrium visible in the α-subunits. In the β-subunits, displacement of the distal valine relative to the haem is necessary in the T-structure before oxygen can bind. In the R-structure this steric hindrance is absent.

2.3.3 *Changes on oxygenation in tertiary structure of the globin*

The α-subunits. Since the $\alpha_1\beta_1$ contact undergoes no significant changes during the R→T transition, the atoms at this contact can serve as a reference frame for changes in tertiary structure elsewhere; except for residues G1–4 and H18–21, the B, G and H helices were also found to be static. The largest movements relative to either of these frames occur in helix F, in segment FG, and in residues G1–4, H18–21 and HC1–3. Fig. 11 shows the haem environment of deoxyhaemoglobin superimposed on that of oxyhaemoglobin. It can be seen that, on oxygenation, helix Fα shifts towards the haem and to the right and carries the FG segment with it. In deoxyhaemoglobin the imidazole of His F8 is tilted

deoxy. The bottom diagram illustrates how the flattening of the porphyrin on going from deoxy to oxy exerts a leverage on leucine FG3 and valine FG5 which lie at the switching contact between the two structures (from Perutz *et al.* 1987). The differences in haem geometry between deoxyhaemoglobins in the T and R structures shown here are closely similar to those found between sterically hindered 2-methyl- and unhindered 1-methylimidazole iron porphyrin complexes synthesized by Momenteau *et al.* (1988) as models of the two states.

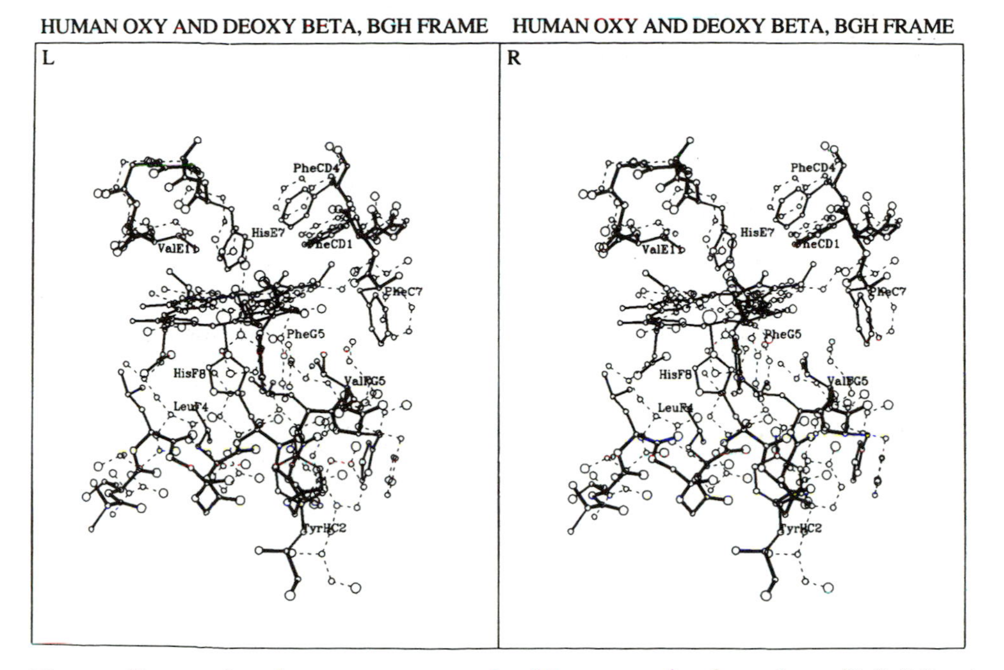

Fig. 12. Change of tertiary structure near the β-haem on going from deoxy-T (full lines) to oxy-R (broken lines). Note how the haem moves to the right, into the haem pocket, and the distal valine and histidine make way for the bound oxygen. (From Perutz *et al.* 1987.)

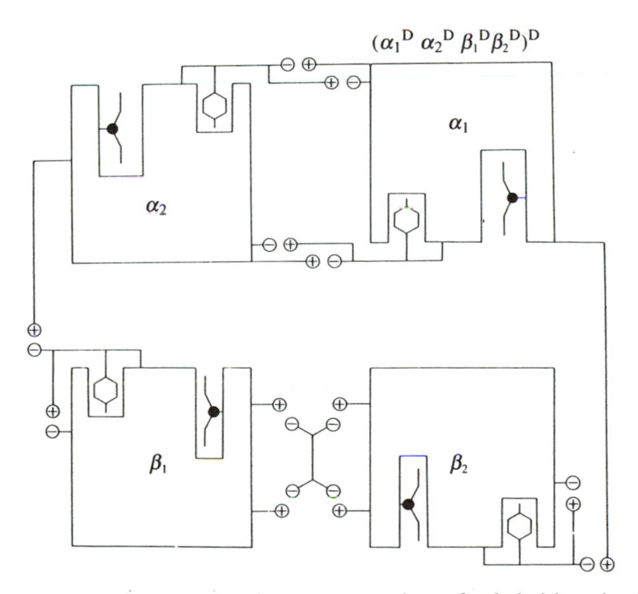

Fig. 13. Diagrammatic representation of salt bridges in the T-structure. Those at the top link the C-terminal Arg HC3(141)α_2 to Asp H9(126)α_1 and Lys H10(127)α_1. The others link the C-terminal His HC3 (146)β_1 to Asp FG1 (94)β_1 and Lys C5(40)α_2. The bridge between the β-subunits represents 2,3-diphosphoglycerate. (From Perutz 1970.)

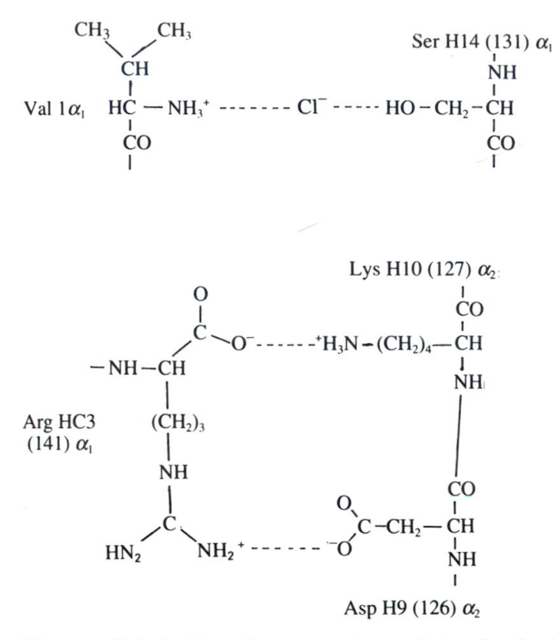

Fig. 14. Salt bridges between the α-chains in the T-structure.

relative to the haem normal; in oxyhaemoglobin the shift of helix F relative to deoxyhaemoglobin aligns it with the haem normal. Taking as a reference frame residues F1–8 to which the haem is attached, the haem flattens and turns clockwise by 10°; the motion of its right-hand edge pushes down Leu FG3(91)α and Val FG5(93)α, which form part of the $\alpha_1\beta_2$ contact where the quaternary switch occurs (Fig. 10). In the T-structure the N- and C-termini form the hydrogen bonds shown in Fig. 13 and 14. In the R-structure these hydrogen bonds are broken, and the terminal residues are seen only at a low level of electron density, implying that they are mobile.

The β-subunits. Fig. 12 shows that on oxygenation helix F moves towards the haem and in the direction of the FG segment, carrying that segment with it and aligning His F8 with the haem normal. The movement of F and FG is transmitted to residue G1 and dissipated beyond G5. The centre of the haem moves further into its pocket along a line linking porphyrin N_1 to N_3, and the haem rotates about an axis close to the line linking N_2 to N_4. Referred to residues F1 to F6, the iron stays still and the porphyrin becomes coplanar with it. In the T-structure $C_\gamma H_3$ of Val E11(67) obstructs the ligand site at the iron; in the oxygenated R-structure that obstruction is cleared by a concerted shift of helices D and E and the CD segment together with the beginning of helix B, away from and across the haem.

The C-terminal histidines form different sets of hydrogen bonds in the T and R structures, as a result of which their pK_a's drop on oxygenation. The conformation of the reactive sulphydryl groups of Cys F9(93)β also changes (Fig. 15) (Perutz *et al.* 1987).

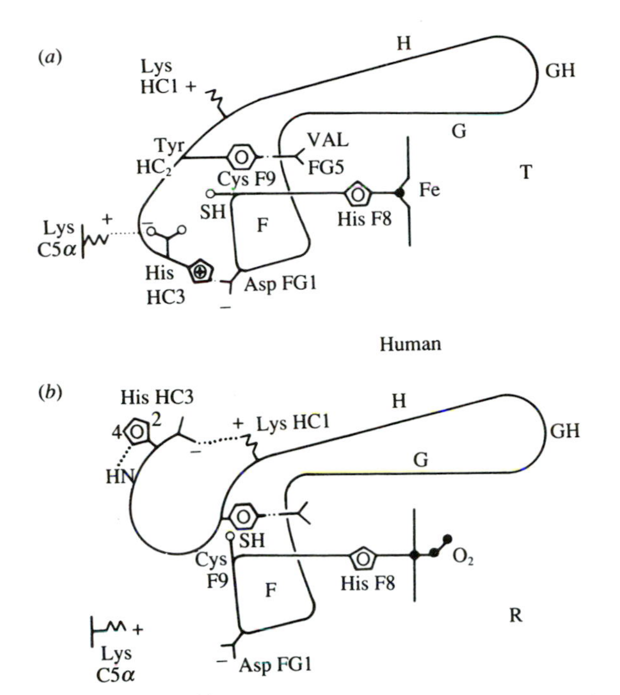

Fig. 15. Change in conformation of histidine HC3 (146)β and cysteine F9 (93)β on going from the T- to the R-structure. In the T-structure the imidazole of the histidine *donates* a hydrogen bond to Asp FG1 and is positively charged ($pK_a = 8{\cdot}0$). Its carboxylate *accepts* a hydrogen bond from Lys C5α. The SH is cis to CO and points away from the haem. In the R-structure the imidazole *accepts* a hydrogen bond from the histidine's main chain NH and has a pK_a of 7·1 or below; and the C-terminal carboxylate accepts a weak hydrogen bond from Lys HC1. The SH group is cis to NH and in contact with Tyr HC2.

In the T-structure the two β-subunits form a binding site for 2,3-diphosphoglycerate (Fig. 16). In the R-structure the gap between the two β-chains becomes too narrow to accommodate it.

2.3.4 *The heterotropic ligands*

According to allosteric theory the low O_2 affinity of the T- as compared to that of the R-structure arises from increased energy and/or number of bonds between the subunits (Monod *et al.* 1965). The contact areas and the number of bonds between segments $C\alpha_1$ and $FG\beta_2$ and between segments $C\beta_2$ and $FG\alpha_1$ are about equal in the R- and T-structures (Baldwin & Chothia, 1979); the C-terminal residues and DPG, on the other hand, form 14 salt bridges between the subunits which are absent in the R-structure (Figs. 13–16). The bond energies of the four pairs of salt bridges made by the C-terminal residues have been measured. Those formed by the C-terminal histidines and histidines H21 (143) of the β-chains together contribute 7·6 kcal mol^{-1} (Louie *et al.* 1988), and those formed by arginine HC3 (141)α contribute at least 4 kcal mol^{-1}, leaving only 300 cal mol^{-1} per salt bridge to be contributed by the remaining eight salt bridges – sufficient to account for the total free energy of cooperativity of 14·4 kcal/tetramer, because a salt bridge

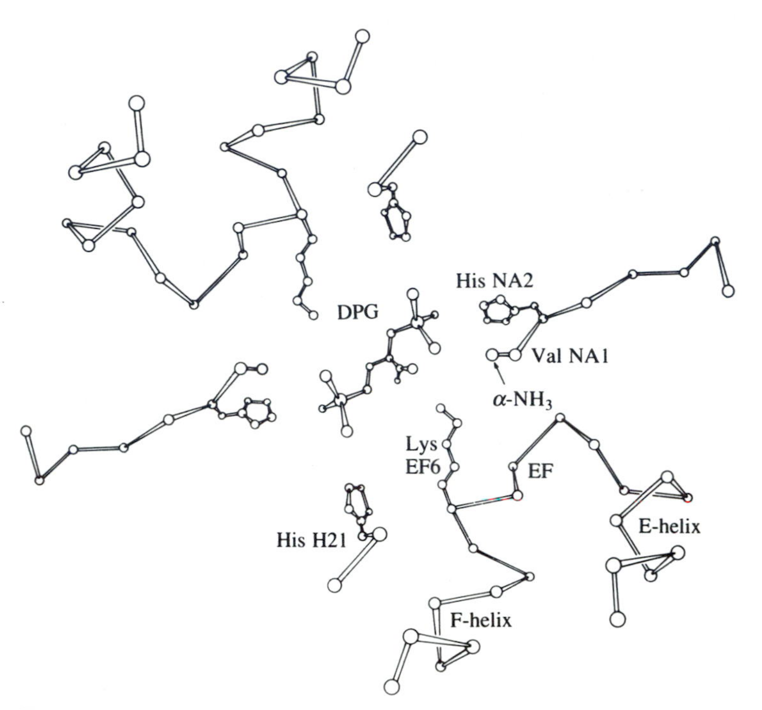

Fig. 16. Hydrogen bonds between 2,3-diphosphoglycerate and cationic groups of the β-chains in the T-structure. In the R-structure the gap between the EF corners closes and the N-termini move apart. (From Arnone, 1972.)

contributes usually at least 1 kcal mol^{-1} (Fersht *et al.* 1986). Absence of any of the bridges raises K_T and lowers L. The salt bridges keep the subunits rigidly in the tertiary deoxy structure and hinder the movement of the iron atoms into the planes of the porphyrin nitrogens and the flattening of the porphyrins themselves. This hindrance manifests itself in spectroscopic and magnetic differences between liganded haemoglobins in the two quaternary structures; these have recently been reviewed (Perutz *et al.* 1987).

All the heterotropic ligands lower the oxygen affinity by forming additional hydrogen bonds that specifically stabilize and constrain the T-structure. The most important heterotropic ligands are protons. The linkage of proton uptake to oxygen release and vice versa, is known as the Bohr effect. For each mole of O_2 released at pH 7·4 and 25 °C, human Hb takes up 0·2 mol H^+ in a deionized solution, 0·5 mol H^+ in 0·1 MCl^-, and 0·7 mol H^+ in the presence of a molar excess of DPG (Kilmartin, 1974; Perutz *et al.* 1980). The identity of the residues that take up protons has been determined by X-ray crystallographic and chemical studies of normal and mutant haemoglobins. In deionized solutions all the protons are taken up by His HC3(146)β, which donates a hydrogen bond to Asp FG1(94)β in the T-structure and accepts a hydrogen bond from its own main chain NH in the R-structure (Fig. 15). In consequence, its pK_a rises from 7·1 or less in oxyhaemoglobin to 8·0 in deoxyhaemoglobin (Kilmartin *et al.* 1973; Matsukawa *et al.* 1984). The binding of Cl^- by the T-structure raises the pK_a's of Val

NA1(1)α (Fig. 14) and Lys EF6(82)β, which contribute an additional 0·28 mol H^+ to the Bohr effect. DPG enters a cleft flanked by the N-termini and helices H of the β-chains and forms hydrogen bonds with Val NA1(1), His NA2(2), Lys EF6(82) and His H21(143) (Fig. 16). The rise in pK_a's of their cationic groups contributes 0·33 mol H^+ to the Bohr effect (Kilmartin, 1974). CO_2 forms carbamino groups with Val NA1(1)α and β and these in turn make hydrogen bonds with cationic groups of the globin. All the groups that bind heterotropic ligands are at some distance from the haems, consistent with Monod *et al.*'s (1963) prediction that 'no direct interaction need occur between the substrate of the protein and the regulatory metabolite which controls its activity'.

In 0·1 M Tris HCl + 0·1 M NaCl at pH 7·4 and 21·5 °C, the first mole of O_2 taken up releases 0·64($\pm$7) mol H^+, the second and third mole of O_2 combined release 1·62($\pm$7) H^+ and the fourth mole of O_2 releases only 0·05($\pm$6) mol H^+ (Chu *et al.* 1984). How is their release related to the allosteric transition from T to R? Allosteric theory allows the equilibrium constant $L_i = [\mathrm{T}]/[\mathrm{R}]$ at the *i*th step of oxygenation to be calculated from $L_i = L_0(K_R/K_T)^i$. Under the above non-physiological conditions $L_1 = 8{\cdot}7 \times 10^4 \times 0{\cdot}0073 = 633$ (Imai, 1982; Baldwin, 1975). Thus more than a quarter of the Bohr protons are discharged before $\frac{1}{600}$ of the Hb molecules have switched from T to R, which implies that the hydrogen bonds responsible for H^+ discharge must break in the T-structure. The bulk of the protons are released in the T$\rightarrow$R transition, which takes place mostly at the second and third oxygenation steps. After the third oxygenation step $L_3 = 0{\cdot}034$, leaving a little more than $\frac{1}{30}$ of the Hb molecules in the T structure, which is roughly equivalent to the fraction of $\frac{1}{20}$ of the protons discharged at the fourth oxygenation step.

2.3.5 *Influence of mutations on the allosteric equilibrium*

Study of the abnormal human haemoglobins has taught us a great deal about the allosteric mechanism. The $\alpha_1\beta_2$ contact acts as a two-way switch between the T and R structures. Each position of the switch is stabilized by a different set of hydrogen bonds. Disruption of any bond that stabilizes specifically the R-structure lowers the oxygen affinity and raises the allosteric constant L, and disruption of any bond that stabilizes the T-structure does the reverse. For example, haemoglobin Kansas (Asn G4(102)$\beta\rightarrow$Thr) has a low oxygen affinity and low Hill's coefficient because the R-structure is destabilized (Fig. 17), while haemoglobin Kempsey (Asp G1(99)$\rightarrow$Asn) has a high oxygen affinity and low Hill's coefficient because the T-structure is destabilized. Hill's coefficient is low in both haemoglobins on account of the bell-shaped curve that relates it to the allosteric constant L. Mutations that disrupt the C-terminal salt bridges in the T-structure also have high oxygen affinities and low Hill's coefficients. The surprise came when a mutation that disrupts hydrogen bonds common to *both* the R and T structures produced the same effect. This happens in haemoglobin Philly (Tyr C1(35)β-$\rightarrow$Phe), where the loss of the phenolic hydroxyl disrupts a network of hydrogen bonds at the $\alpha_1\beta_1$ contact which does not change in the T$\rightarrow$R transition. It has since become clear that the loss of any bond, either between or within the subunits, even the creation of a cavity, relaxes the T-structure. This relaxation

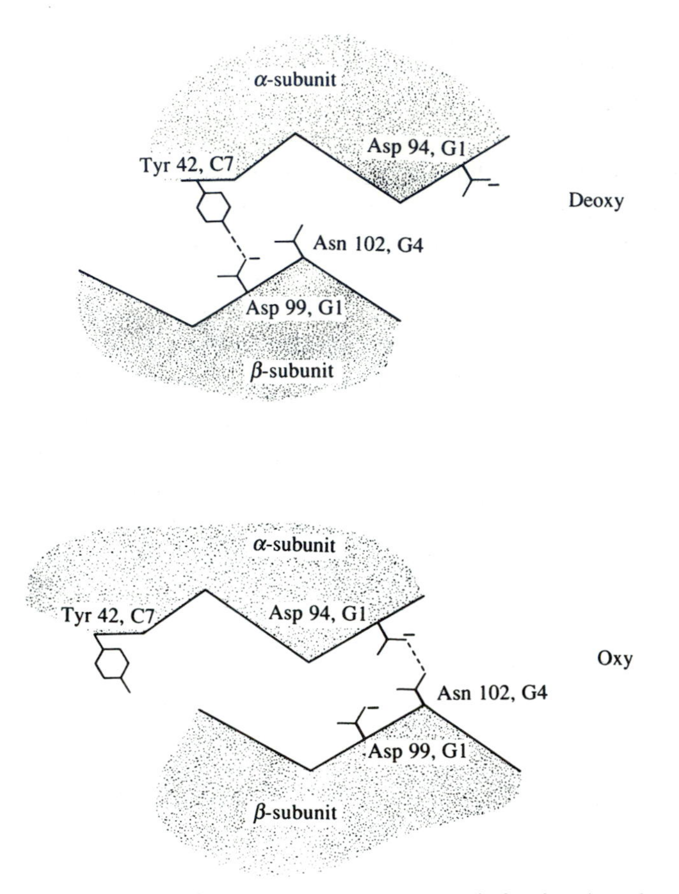

Fig. 17. The $\alpha_1\beta_2$ contact as a two-way switch, showing alternative hydrogen bonds stabilizing the deoxy-T and oxy-R structures.

raises K_T and lowers L. K_R is unaffected, since it is almost the same as that of free α- and β-subunits or $\alpha\beta$ dimers. In all instances changes in L are linked to changes in K_T; it is not possible to alter these parameters independently (Fermi & Perutz, 1981).

2.3.6 *Role of the distal residues and molecular dynamics*

Despite the enormous amount of research done on the structure and function of haemoglobin, some of its most vital properties have remained ill-understood. Free ferrous porphyrins are rapidly oxidized by oxygen, and their affinity for oxygen is several thousand times smaller than that for carbon monoxide. Globin keeps the iron ferrous, which is necessary because only ferrous iron combines reversibly with molecular oxygen; globin also discriminates in favour of oxygen and against carbon monoxide. This is essential for life, since carbon monoxide is produced endogenously in the breakdown of porphyrin that follows the lysis of red cells. A combination of directed mutagenesis and functional studies has recently clarified these problems.

In myoglobin and in the α-subunits of haemoglobin, N_ϵ of the distal histidine forms a hydrogen bond with the bound oxygen, but not with CO; in the β-subunits that bond is either weak or absent (Phillips & Schoenborn, 1981; Shaanan, 1983; Cheng & Schoenborn, 1989). X-ray analysis shows that the distal histidine blocks access to the haem pocket (Perutz & Mathews, 1966). Neither O_2 nor CO can enter or leave unless the side-chain of the distal histidine swings out of the way, which it can do only by elbowing the helix E, to which it is attached, away from the haem. Thus oxygen transport relies on the dynamics of the globin.

Olson *et al.* (1988) have replaced the distal histidines in myoglobin and in the α and β subunits of haemoglobin by glycines, which opened access to the haem pockets, and have measured the resulting changes in the rates of association with and dissociation from O_2, CO and a more bulky ligand, methylisocyanide. Their results are best analysed in terms of transition-state theory (Szabo, 1978). If the transition state is product-like, any rise in affinity can be brought about mainly by a rise in the rate of association. If it is reactant-like, any rise in affinity can be achieved mainly by a drop in the rate of dissociation. These rules generally hold, even though neither rate gives directly the rate of formation or dissociation of the transition state.

The replacement of the distal histidine by glycine leaves the oxygen affinity and kinetic constants of the β-subunits unchanged within error; it *diminishes* the oxygen affinity of the α-subunits 8-fold, equivalent to stabilization of the bound oxygen by hydrogen bonds with the histidines by the equivalents of 1·0 kcl mol^{-1}. The reduction in affinity is brought about by a 60-fold increase in the dissociation (*off*) rate that more than compensates the 10-fold increase in association (*on*) rate due to the opening of the haem pocket. The acceleration brings the on-rate to a value of $15(\pm 5) \times 10^7\ s^{-1}\ M^{-1}$ which is close to Szabo's estimate of $50 \times 10^7\ s^{-1}\ M^{-1}$ for a hypothetical globin in which that rate is limited only by diffusion into the haem pocket through a hole of 2·6 Å radius. The absence of any acceleration by the His → Gly replacement in the β-subunit implies that in native haemoglobin histidine E7β must be swinging in and out at least 10^9 times per second, while in myoglobin and in the α-subunits that rate appears to be about a hundred times slower. All the *off* rates are several orders of magnitude slower than the *on* rates, which indicates that the rate-limiting step for the *off* rates is rupture of the Fe—O bond rather than opening of the haem pocket.

Replacement of the distal histidine by glycine *increases* the affinity for carbon monoxide 4-fold in the α subunits, due largely to a rise in the *on* rates, and *decreases* it 3-fold in the β subunits. In the native proteins, the *on* rates for CO are slower by an order of magnitude than those for oxygen, whence the acceleration is likely to be due to the removal of static steric hindrance by the distal histidines within the haem pocket rather than its function as a gate. If we multiply the decrease in oxygen affinity by the increase in affinity for CO, we find that the distal histidine discriminates against CO by the equivalent of about 2 kcal mol^{-1} in the α subunits, but does not discriminate in the β-subunits.

The switch in quaternary structure from R to T reduces the oxygen affinity of haemoglobin by the equivalent of over 3 kcal per mol haem. Like the reduction in oxygen affinity due to the His → Gly replacement, it is brought about mainly by acceleration of the *off*-rates. Hence all the evidence points to the transition state with oxygen being mainly reactant-like. By contrast, the decrease in CO affinity in the R → T transition is due to a drop in the *on* rates, consistent with the present evidence that transition state is mainly product-like.

What role does the distal valine play in the discrimination between O_2 and CO? Its replacement by alanine increases both the *on* and *off* rates of oxygen with the α-subunits 7-fold and leaves those with the β-subunits unchanged. It increases the *on* rate of CO with the α-subunits 10-fold and leaves the *off* rates unchanged. Hence the distal valine in the α-subunits discriminates against CO by the equivalent of 1·3 kcal mol^{-1}, apparently by steric hindrance, but is ineffective in the β-subunits. However, this is only part of the story, because the results of Olson *et al.* (1988) are confined to the R-structure. In the T-structure steric hindrance by the distal valine E11β plays a key role, which future experiments may quantify. Nevertheless, the mechanism of discrimination in the β-subunits remains a mystery.

Other evidence on the mechanism of discrimination between O_2 and CO has come from recent crystal structure determinations. In synthetic model compounds that offer no steric hindrance to the ligands, oxygen binds with an Fe—O—O angle of 120°, while CO lies on the haem axis. The haem pockets of myoglobin and haemoglobin seemed to be tailored to accommodate the bent oxygen and force the CO off the haem axis, which might have accounted for their low CO affinity. This is true in myoglobin, where CO is seen in two orientations, with Fe—C—O inclined at either 120° or 140° to the haem axis (Kuriyan *et al.* 1986). On the other hand, recent X-ray analyses of human carbonmonoxyhaemoglobins at 2·2–2·3 Å resolution have shown inclinations of less than 10° – too little to account for the observed energy of discrimination. There are no significant displacements of the distal residues, but the porphyrins are ruffled (Derewenda *et al.* 1989). X-ray analysis of a synthetic 'hindered pocket' iron porphyrin that has a CO affinity lower than that of the unhindered 'picket fence' iron porphyrin by the equivalent of 1·2 kcal/mol shows similar geometry. Fe—C—O is inclined to the haem axis by only 7·5° and the porphyrin is markedly ruffled (Kim *et al.* 1989). It looks as though in both the 'hindered pocket' porphyrin and in haemoglobin a major part of the strain energy responsible for the low CO affinity may be stored in the porphyrin.

According to the atomic models, replacement of valine E11β by isoleucine should block the oxygen site in both the T and R structures, but in fact isoleucine at E11 fails to inhibit oxygen binding: it merely shifts the entire equilibrium curve to the right, roughly halving the oxygen affinity, which implies a fourfold reduction of the affinity of the β-haems if the α-haems remain unaffected. It appears that both the T and R structures have enough flexibility to adjust the relative positions of the haem and the distal isoleucine sufficiently for ligands to bind, at a cost of only about 1·6 kcal/mole β-haem (Matthews *et al.* 1989).

Computer simulations of the molecular dynamics of the exit of carbon monoxide from the interior of myoglobin suggest that there may be several alternative pathways in addition to that via the distal histidine (Elber & Karplus,

1989), even though the latter is the most direct. Experimental evidence in support of the histidine as the door to the haem pocket comes from the structure of imidazole and phenylhydrazine myoglobin in which the side-chain of the distal histidine has been turned out of the haem pocket by the bulk of the ligand (Bolognesi *et al.* 1982; Ringe *et al.* 1984) and from a recent crystal structure determination of ethylisocyanidemyoglobin that shows the side-chain of the distal histidine in two alternative positions, either in or out of the haem pocket, exactly as it would have to move to admit or release ligands (Johnson *et al.* 1989). The dynamic movements of the haem pocket are attested by NMR studies showing that phenylalanines CD1 and CD4, which wedge the haem into its pocket and are packed tightly between the haem and the distal helix E, flip over at rates faster than 10^4 s^{-1} (Dalvit & Wright, 1987). They can do so only if the entire haem pocket breathes fast. The fast exchange of most main-chain imino hydrogens with tritium also attests to the dynamic state of the haemoglobin molecule (Englander & Kallenbach, 1984). Oxygen has been found to quench the fluorescence of buried tryptophans in other close-packed proteins at velocities near the diffusion limit, which could not have happened unless the proteins' dynamic motion had opened gaps wide enough to let the oxygen pass (Lakowicz & Weber, 1973; Calhoun *et al.* 1983).

Springer *et al.* (1989) have studied the protection of the haem iron from oxidation by replacing the distal histidine in sperm-whale myoglobin by ten different amino acid residues and shaking the deoxygenated myoglobin solutions in air in 75 mM potassium phosphate + 25 mM EDTA pH 7·0 at 37 °C. All replacements reduced the oxygen affinity and accelerated autoxidation. Phenylalanine, methionine and arginine produced the smallest accelerations (~ 50-fold); aspartate the largest (350-fold).

How can these results be interpreted? Paradoxically, combination with oxygen protects the haem iron from oxidation, as can be shown by performing the same experiments at several atmospheres of pure oxygen. Apparently oxidation occurs in that fraction of molecules which are deoxygenated at any one moment. The larger that fraction is at atmospheric oxygen pressure, the faster myoglobin autoxidizes. For example, replacement of the distal histidine by phenylalanine reduced the oxygen affinity 170-fold, so that a larger fraction of myoglobin molecules will have remained deoxygenated at atmospheric oxygen pressure and therefore have become autoxidized. However, this can be only part of the explanation, because the replacement of histidine by glycine reduces the oxygen affinity merely 11-fold, yet accelerates autoxidation over 100-fold.

Autoxidation is catalysed by protons, hence the 350-fold acceleration by aspartate. I suggest that the distal histidine protects the ferrous haem iron by acting as a proton trap. The distal histidine has a pK_a of about 5·5; at neutral pH it is protonated only at N_δ which faces the solvent. Any proton entering the haem pocket of deoxymyoglobin would be bound by N_ϵ, and simultaneously N_δ would release its proton to the solvent. When the histidine side-chain swings out of the haem pocket, the protons would interchange, restoring the previous state. No other amino acid side-chain could function in this way. Evolution was a brilliant chemist.

2.3.7 *Drugs as allosteric effectors*

In a search for compounds that might prevent the aggregation of deoxyhaemoglobin S in patients with sickle-cell anaemia, two antilipidaemic drugs, clofibric acid and its analogue bezafibrate, were found to lower the oxygen affinity of haemoglobin. X-ray analysis of crystals grown in the presence of these compounds showed that they combine with deoxy-, but not with oxyhaemoglobin. They stabilize the T-structure by combining with sites in the central cavity that are about 20 Å away from the DPG binding sites; their effects and that of DPG on the allosteric equilibrium are additive (Perutz *et al.* 1986).

This discovery led I. Lalezari to synthesize a family of new compounds related to, but more active than bezafibrate (Lalezari *et al.* 1988); one of these, LR30, which has the formula shown below, has turned out to be the most powerful allosteric effector yet found (Fig. 18). At an effector concentration equimolar to haem and at pH 6, it lowers the oxygen affinity of solutions of human haemoglobin to a level found until now only in fish haemoglobins that exhibit a Root effect. In such fish haemoglobins, the tension on the haem in the T-structure that is associated with the low oxygen affinity has been demonstrated directly.

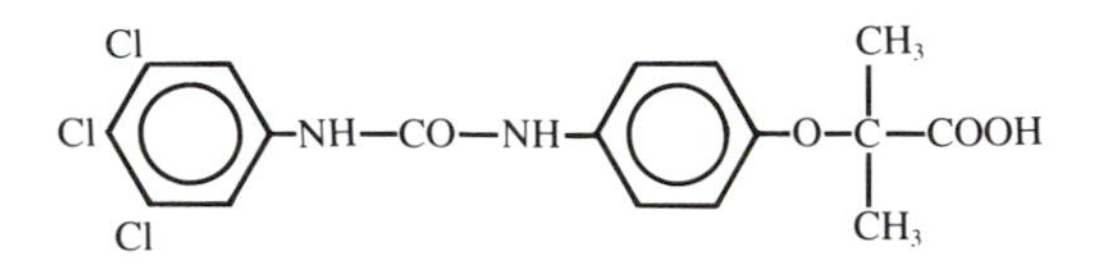

In azidemethaemoglobin the iron is in a thermal equilibrium between two different spin states that are characterized by different lengths of the iron nitrogen bonds; they are longer in the high spin than in the low spin state. In azidemethaemoglobin of trout, transition from the R to the T structure induces a transition to higher spin, equivalent to a stretching of the Fe—N bonds (Perutz *et al.* 1978; Messana *et al.* 1978). This is manifested by an increase in paramagnetic susceptibility equivalent to a change in free energy of 1 kcal/mol (Fig. 19) and also by a change in optical spectra which shows a rise in intensity of the high-spin bands at 500 and 630 nm and a drop in intensity of the low-spin bands at 540 and 570 nm. Until recently it was not possible to induce this transition in human azidemethaemoglobin (Philo & Dreyer, 1985), but combination with the powerful new synthetic effector induced a difference spectrum identical to that of trout haemoglobin which demonstrates that the haems are under tension also in human haemoglobin in the liganded T-state (Fig. 20).

X-ray analysis and oxygen equilibria show that four molecules of the effector

---

Fig. 18. (*a*) Novel allosteric effectors of haemoglobin: bezafibrate, an antilipidaemic drug, and compounds derived from it by Dr I. Lalezari. (*b*) Effect on partial pressure of oxygen at half saturation (P_{50}) of 2,3-diphosphoglycerate (DPG), inositol hexaphosphate (IHP) and two of Dr Lalezari's synthetic effectors LR20 and LR30. The inset shows their effect on Hill's coefficient at half-saturation. At 1 mM LR30 the haemoglobin is only half-saturated at 2 × atmospheric oxygen pressure and its Hill's coefficient is near unity because the allosteric constant L is very large. (Courtesy Dr C. Poyart.)

(*a*) Bezafibrate

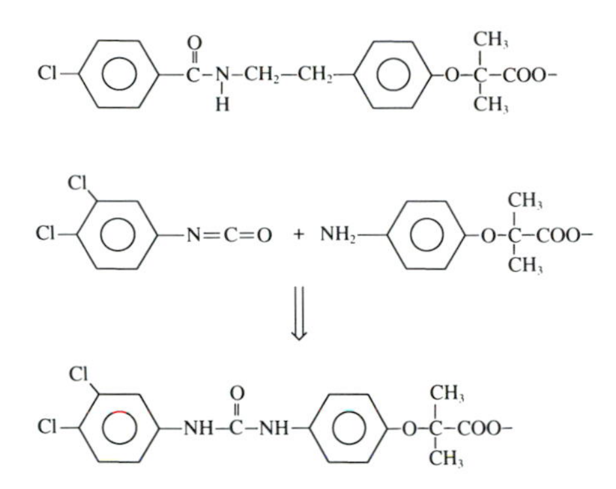

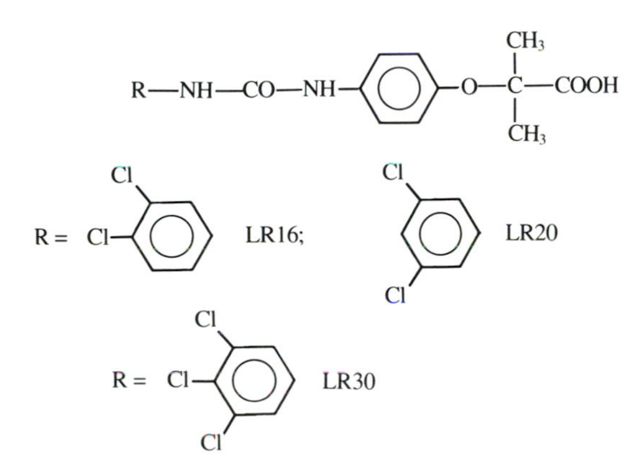

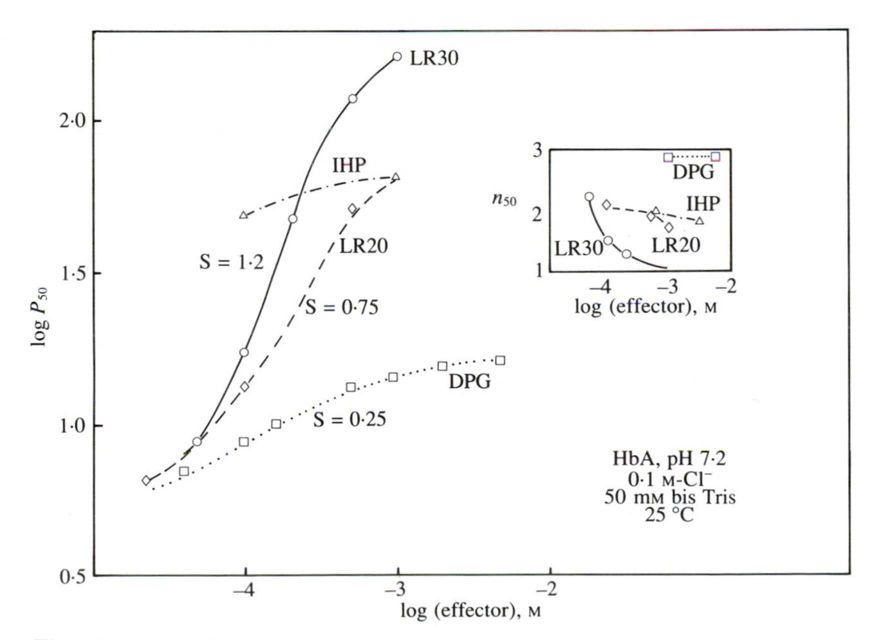

Fig. 18 (*a* & *b*). For legend see opposite.

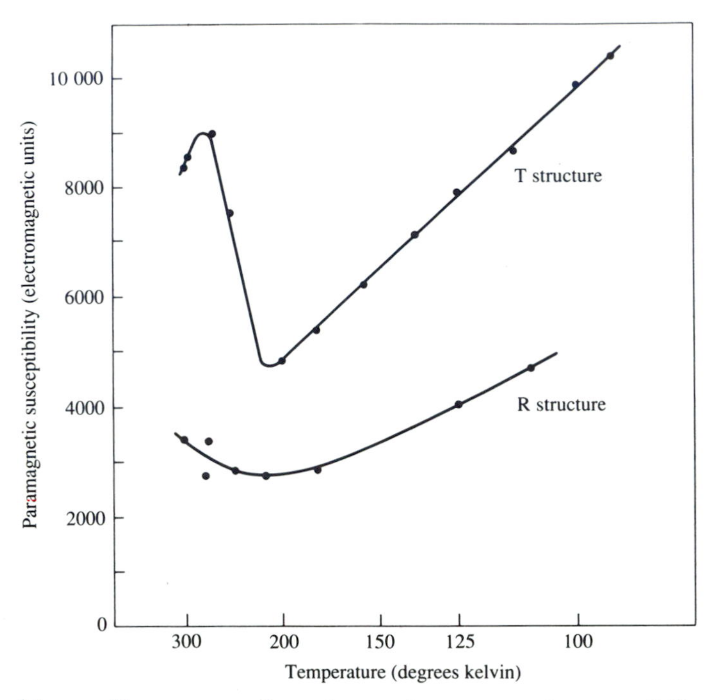

Fig. 19. Temperature dependence of paramagnetic susceptibility of carp azide methaemoglobin in the R- and T-structures. Below about 200 °K the susceptibility rises with falling temperature in accordance with Curie's law. Above that temperature a thermal equilibrium between a high and a low spin form masks that behaviour, since the high spin form gains stability with rising temperature. It has longer Fe—N bond distances than the low spin form. Tension at the haem in the T-structure therefore shifts the equilibrium towards higher spin. (From Messana *et al.* 1978.)

combine with one molecule of human deoxyhaemoglobin both in the crystal and in solution. One pair of symmetry-related binding sites is close to those of bezafibrate; the other pair is nearly at right angles to the first (Fig. 21). The four trichlorobenzene moieties form a parallel stack. Their close packing seems to dominate their mode of binding. The combination of human deoxyhaemoglobin with the effectors induces no stereochemical changes at the allosteric core that are visible at 2·5 Å resolution and produces a decrease of the Fe—N_ϵ stretching frequency by only 1–2 cm^{-1} (Kitagawa 1989). Only in the liganded T-structure do the effectors make themselves felt at the haems.

The work on drug binding by haemoglobin has led to generalizations that are relevant to the binding of effectors, transmitters and drugs to other proteins. The stereochemistry of binding is determined by the available van der Waals space and, within that space, by interactions of a wide range of polarity, from strong hydrogen bonds between ionized groups of opposite charge to weak interactions between aromatic quadrupoles, and non-polar interactions between aliphatic hydrocarbons. The detailed streochemistry is governed by a tendency to maximize the sum of the energy of electrostatic interactions, for example by aligning effectors relative to the protein so that the mutual polarizabilities are maximized. Drugs

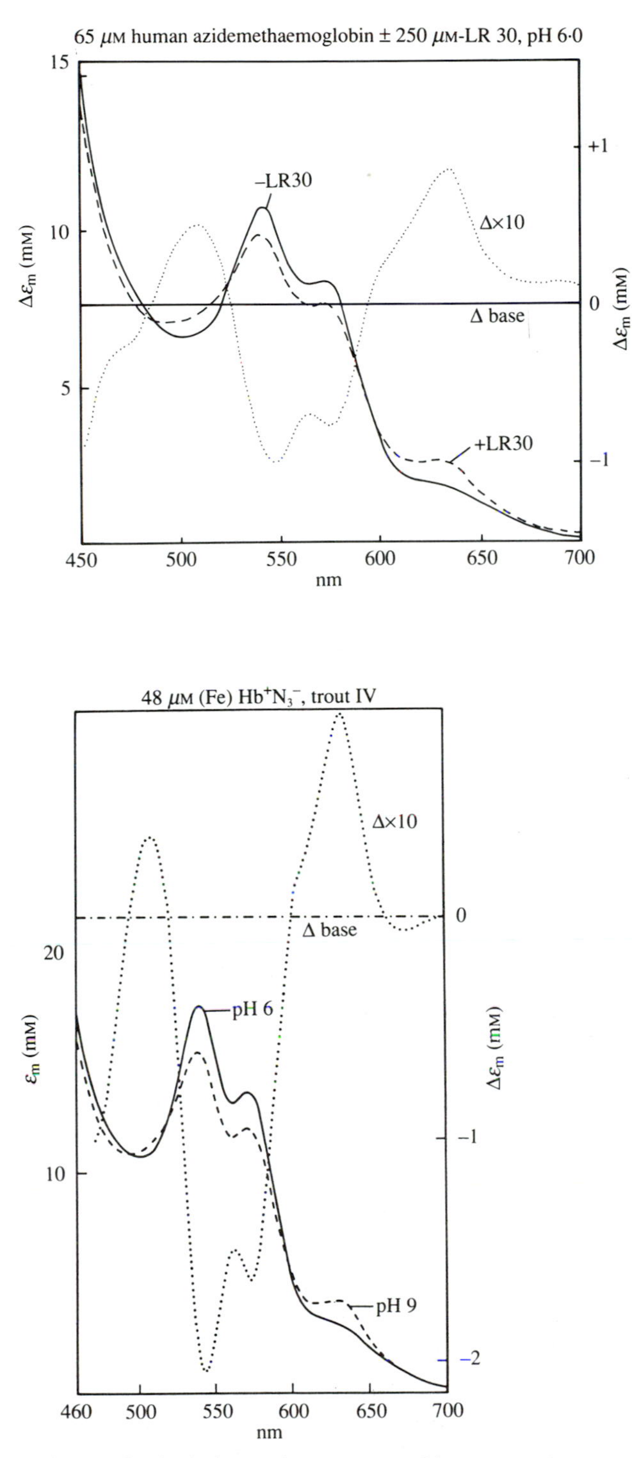

Fig. 20. Optical absorption spectra of human and trout IV azide methaemoglobins in the R- and T-structures, showing the rise in intensity of the high spin bands at 500 and 630 nm and the fall of the low spin bands at 540 and 570 nm. Full lines, T-structure; broken lines, R-structure; dotted lines, difference spectra; 10 × enlarged.

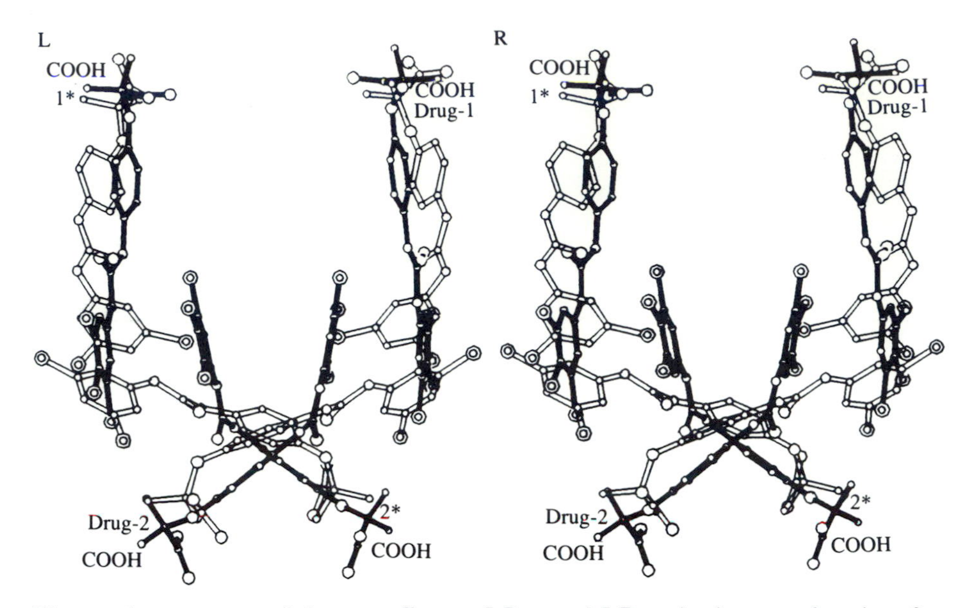

Fig. 21. Arrangement of the two effectors LR20 and LR 30 in the central cavity of haemoglobin. The vertical pairs lie mainly between the two α-chains in sites that overlap those of bezafibrate. The other pairs occupy positions not taken up by bezafibrate, and the sites of the 3,5-dichloro derivative (LR20) differ from those occupied by the 3,4,5-trichloro derivative (LR30). The four trichlorobenzene moieties form a close-packed parallel stack right in the centre of the haemoglobin molecule, while the dichlorobenzene moieties stack in separate pairs. (G. Fermi & M. F. Perutuz, unpublished.)

may influence the allosteric equilibrium of a protein receptor in the same direction as the natural effector even though they are chemically unrelated to it, because proteins may offer a variety of binding sites not used in nature (Perutz *et al.* 1986).

3. HAEMOCYANIN: DEPENDENCE OF ALLOSTERIC EQUILIBRIUM ON COORDINATION AND VALENCY OF A BINUCLEAR COPPER COMPLEX

Oxygen carriers of invertebrates are metallo-proteins that are freely dissolved in the body fluids rather than sequestered in specialized cells. Several of these proteins have very large molecular weights, perhaps to prevent their passage through cell membranes. The respiratory carriers of arthropods and molluscs are copper-proteins known as haemocyanins, because they are blue when oxygenated. In the absence of oxygen they are colourless. Their reaction with oxygen is cooperative and accompanied by a release of protons like that of haemoglobin, but calcium or sodium, as well as chloride ions, act as allosteric inhibitors, while lactate can be an activator (Johnson *et al.* 1984). Haemocyanins are oligomers which can be dissociated into their component subunits. They also combine with carbon monoxide, but for reasons explained below, and unlike haemoglobin, that reaction is non-cooperative (Bonaventura *et al.* 1973; van Holde & Miller, 1982; Ellerton *et al.* 1983).

Molluscan haemocyanins form hollow cylindrical assemblies of 10–20 subunits

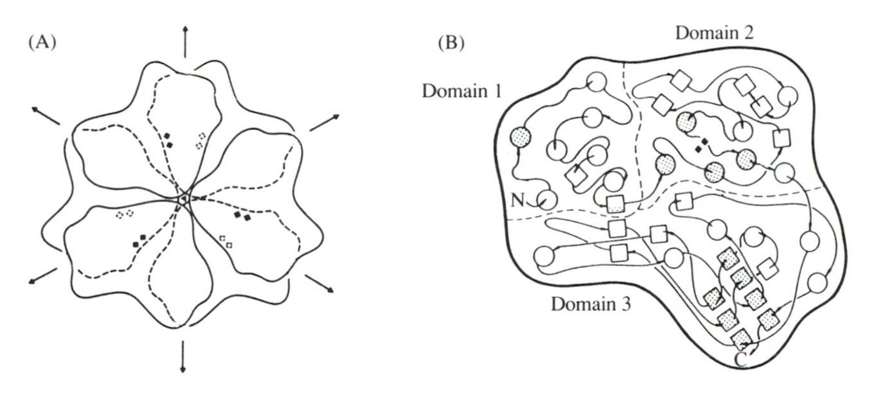

Fig. 22. (A) Quaternary structure of haemocyanin of the spiny lobster, showing the outlines of the two trimers stacked face to face. The arrows show the dyad symmetry axes of the point group D_3 or 32. The full squares are copper atoms in the top trimer, the empty squares copper atoms in the bottom trimer. (B) Tertiary structure of subunit, indicating α-helices as circles, β-strands as squares and copper atoms as small squares. (From Linzen et al 1985.)

with molecular weights of 3·5–9 million. For example, the subunit of *Octopus dofleini* haemocyanin has a molecular weight of 350 000 and consists of seven homologous domains in tandem, each containing a binuclear copper site. Ten such subunits are assembled in the complete molecule (Lamy *et al.* 1987). The isolated domains bind oxygen non-cooperatively, but the complete molecule binds it with a Hill's coefficient greater than 3 and exhibits a Bohr effect $\Delta \log P_{50}/\Delta$pH of $-1{\cdot}7$, more than 3 times that of haemoglobin in 0·1 M NaCl without DPG. A family of oxygen equilibrium curves between pH 6·6 and 8·0 looks like that of haemoglobin in Fig. 5, p. 5, except that at the extremes of pH the curves come even closer to being straight lines at 45° to the axes. The changes in quaternary structure responsible for the strong cooperative effects are still unknown (Miller, 1985).

Arthropod haemocyanins are hexamers composed of subunits of MW 75 000, each containing two atoms of copper. W. Hol and his colleagues have determined the structure of deoxyhaemocyanin of the spiny lobster (*Panulirus interruptus*) at 3·2 Å resolution (Gaykema *et al.* 1984; Linzen *et al.* 1985; Volbeda & Hol, 1986*b*). It consists of six subunits which have a higher oxygen affinity in isolation than when assembled into a hexamer. Ca^{2+} lowers the oxygen affinity (Kuiper *et al.* 1975, 1979). The subunits from three symmetrical dimers are arranged around an axis of threefold symmetry, rather like the catalytic subunits of aspartate transcarbamylase described in Section 7 below. Each subunit contains a single polypeptide chain of 657 amino acid residues, folded into three domains of complex structure (Fig. 22). The central domain contains the copper binding site in the middle of four α helices. Each copper ion is coordinated tightly to the N_ϵ's of two histidines and more loosely to a third. Two of the histidines are separated by three residues on one helix and the third is placed on another helix that is antiparallel to the first. The three N_ϵ's form the base of a trigonal pyramid with the copper at its apex (Fig. 23). The two copper ions are 3.7 Å apart in deoxy and

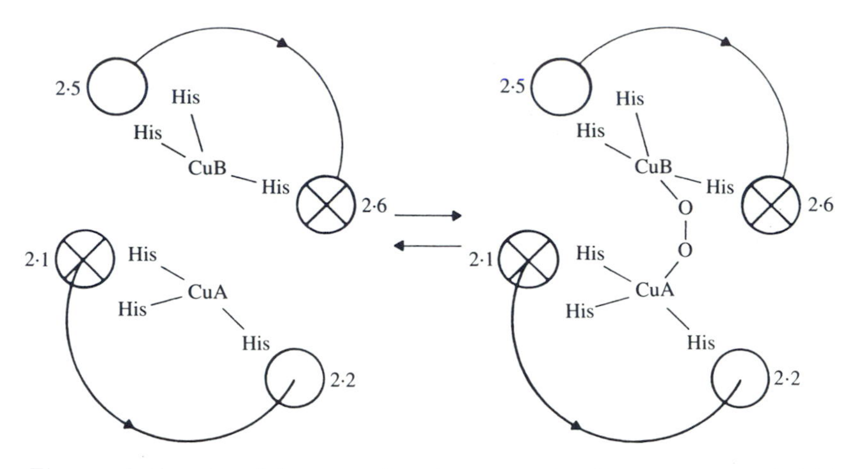

Fig. 23. Active site of deoxy- and oxyhaemocyanin. An oxygen atom bridging the two copper atoms in deoxyhaemocyanin was inferred from spectroscopic evidence, but the refined electron-density map does not show it. The circles represent α-helices. The crosses indicate that the amino ends of the α-helices face the observer. (From Volbeda & Hol, 1989*a*.)

3.4 Å apart in oxyhaemocyanin (Brown *et al.* 1980). The electron-density map shows no bridging atom between them.

In deoxyhaemocyanin the copper ions are cuprous. Combination with oxygen is accompanied by formal oxidation to cupric copper, and the oxygen becomes peroxide-like, as shown by the O—O stretching frequencies of 744 or 749 cm^{-1}, compared to 1107 cm^{-1} in oxyhaemoglobin which is close to that of a superoxide ion, and much lower than that of an oxygen molecule (1556 cm^{-1}). The peroxide ion is believed to bridge the two copper ions. Cu^{+} is a d^{10} ion in which all five *d* orbitals are filled and it is therefore diamagnetic. When combined with four nitrogenous ligands it is most stable in tetrahedral coordination, where the Cu—N bonds bisect the axes of the d-orbitals and repulsion between the copper and nitrogen orbitals is therefore minimized. Cu^{2+}, on the other hand, is normally surrounded by four nitrogenous ligands arranged at the corners of a square. In haemocyanin, the arrangement of the three histidine N_ϵ's at the base of a pyramid with copper at its apex stabilizes the cuprous state. In addition, an electrostatic factor helps to make the reaction of the cuprous ions with oxygen reversible. The active site is lined with hydrophobic side-chains, and there is only one anionic residue, a glutamate at a distance of 7 Å from the copper ions, to compensate their positive charges. In the cuprous state there would be one uncompensated positive charge, but electronic oxidation to the cupric state as in methaemocyanin would create three uncompensated positive charges in a largely hydrophobic cavity, resulting in a large rise in free energy.

The source of the cooperativity of the reaction of haemocyanin with oxygen must be sought in stereochemical changes at the active site. Cupric ion has a spin of $S = \frac{1}{2}$, but in oxyhaemocyanin the spins of the two cupric ions are antiferromagnetically coupled, so that the spin transition should not produce a stereochemical change. On the other hand, the ionic radius of copper shrinks from

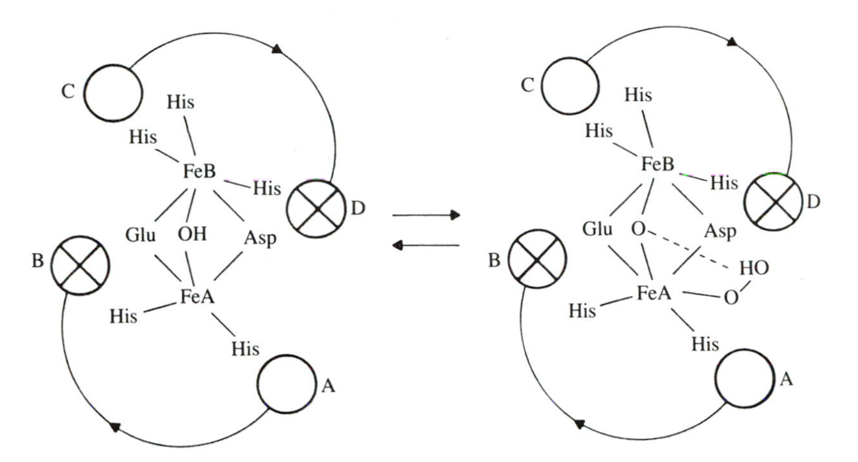

Fig. 24. Active site of deoxy- and oxyhaemerythrin. (From Volbeda & Hol, 1989.)

0·95 Å in cuprous to 0·72 Å in cupric ion, and the distance between the copper ions shrinks by 0·2–0·3 Å on oxygenation (Brown *et al.* 1980). This indicates that on oxygenation the distance between opposite helices in the active site may shrink by 0·6–0·7 Å, comparable to the shrinkage of 0·55–0·60 Å between the proximal histidine and the porphyrin in haemoglobin. This shrinkage is likely to trigger the transition from the quaternary T to the R structure. The low oxygen affinity of the T-structure is probably due chiefly to restraints by the protein that oppose the contraction of the copper complex needed for oxygenation. Conceivably electrostatic effects, such as the motion of ionized residues relative to the active site, could also contribute, but the tight packing of the four α-helices in the centre of the molecule offers little elbow room for such movements, and besides, the reaction with oxygen involves no change in net charge.

The monomers forming the dimer are rigidly linked, which suggests that the allosteric transition may consist of a rotation of the dimers around the twofold axes, coupled with a translation along them. The resulting changes at the subunit contacts are likely to be responsible for the Bohr effect and the altered affinity for Ca^{2+} and Na^{+}. Combination with oxygen occurs at a rate of 3×10^{7} $\text{M}^{-1}\,\text{s}^{-1}$, very near to the maximum rates observed in haemoglobin, but the structure shows no ready access to the copper sites, which suggests that oxygen diffusion relies on protein dynamics. Oxygen dissociation is much slower (60 s^{-1}) (Kuiper *et al.* 1977).

Carbon monoxide binds to only one of the copper ions of each pair. That reaction cannot induce significant stereochemical changes in the copper complex and is therefore non-cooperative (Brunori *et al.* 1982).

4. HAEMERYTHRIN: COOPERATIVITY IN A BINUCLEAR IRON COMPLEX

It is interesting to compare haemocyanin to another oxygen carrier of invertebrates, haemerythrin. This is found in muscles as a monomer of MW

13 000, called myohaemerythrin, and in blood as an octamer of myohaemerythrin-like subunits. Most octameric species were found to bind oxygen non-cooperatively without a Bohr effect, but the haemerythrin of a brachiopod exhibited cooperativity, with a Hill's coefficient of 2·0 (low for an octamer), a free energy of cooperativity of 1·5 kcal mol^{-1}, and a Bohr effect (Manwell, 1960; Richardson *et al.* 1983). Apart from these two publications, all studies have concentrated either on myohaemerythrin or on non-cooperative octamers.

The active site of haemerythrin consists of a pair of iron atoms linked to four α-helices, like the copper-binding α-helices in haemocyanin (Fig. 24). The iron atoms are coordinated to five histidines and are bridged by the carboxylates of a glutamate and an aspartate, such that one iron is five- and the other six-coordinated (Stencamp *et al.* 1985; Sheriff *et al.* 1987). In oxy-, hydroxymet- or azidemethaemerythrins the irons are also linked by a μ-oxo bridge which manifested itself by its Fe—O—Fe stretching frequency of 486 cm^{-1}, and is confirmed by a peak of electron density between the iron atoms in the crystal structures. Replacement of $H_2{}^{16}O$ as a solvent by $H_2{}^{18}O$ reduced that frequency by 9 cm^{-1}, which showed that the bridging oxygen is exchangeable (Shiemke *et al.* 1984; Reem & Solomon, 1987) Both azide ion and oxygen bind end-on to the five-coordinated iron, as in haemoglobin and unlike haemocyanin (Sheriff *et al.* 1987; Stencamp *et al.* 1985). ν(O—O) is 844 cm^{-1}, still in the peroxide range, but 100 cm^{-1} higher than in haemocyanin, because the peroxide ion is bound to only one metal, which makes the O—O bond stronger. $\nu(Fe—O_2)$ in oxyhaemerythrin is 503 cm^{-1}, compared to 567 in oxyhaemoglobin, showing that the Fe—O bond is weaker in haemerythrin. One reason for this weakening became apparent when H_2O was replaced by D_2O as a solvent. ν(O—O) at 844 cm^{-1} increased by 4 cm^{-1} and $\nu(Fe—O_2)$ at 503 cm^{-1} decreased by 3 cm^{-1}, showing that a proton bound to the oxygen must have been replaced by a deuterium. Reem & Solomon have suggested that oxygen is bound to the iron as an $HO_2{}^-$ anion and that this donates a hydrogen bond to the μ-oxo bridge (Fig. 24). They found no such isotope shift in the N—N stretching frequency of azidemethaemerythrin.

Mössbauer spectra show that combination with oxygen is accompanied by oxidation of both iron atoms from the ferrous to the ferric state. The iron atoms remain high-spin, and their spins are antiferro-magnetically coupled, presumably via the bridging oxygen. The coupling is strong in oxy- and weak in deoxyhaemerythrin. Magnetic circular dichroism indicates that in deoxyhaemerythrin the bridge is formed by a hydroxyl ion rather than an oxygen atom; on oxygenation the proton would merely shift from the hydroxyl to the peroxo ion, as shown in Fig. 24, so that there would be no Bohr effect at the active site (Reem & Solomon, 1987).

How could such a complex give rise to cooperative oxygen binding? The ionic radius of high spin Fe^{2+} is greater by 0·16 Å than the radius of high spin Fe^{3+}. That expansion would stretch the distance between helices A and C by 0·64 Å, and that between helices B and D by 0·32 Å, sufficient to trigger an allosteric transition from an R- to a T-structure.

The most reliable pointer to a possible allosteric trigger comes from crystal structures of binuclear ferric iron complexes synthesized as models of haemerythrin by S. J. Lippard and his associates (Armstrong & Lippard, 1984; Armstrong *et al.* 1984). In these complexes each of the iron atoms is coordinated to three nitrogens, mimicking the histidines in haemerythrin; the iron atoms are bridged by the carboxylates of two acetates, mimicking the asparate and glutamate in haemerythrin; finally they are bridged by an oxygen atom, as seen in the refined crystal structures of azide and hydroxymethaemerythrins. In one of the compounds oxygen forms a μ-oxo bridge; in the other it is protonated and forms a hydroxyl bridge. Protonation lengthens the Fe—O distance from an average 1·784 to one of 1·956 Å, and the Fe—Fe distance from 3·146 to 3·439 Å. The shorter distances would correspond to oxy- and the longer ones to deoxyhaemerythrin. The latter correspondence does not strictly apply, because deoxyhaemerythrin is ferrous and the model complex is ferric, and we have seen that oxidation alone may shorten the Fe—Fe distance by about 0·3 Å. Also one of the irons in deoxyhaemerythrin is 5-coordinated, while in the model compound both are six-coordinated; it is not clear what difference that makes. Zhang *et al.* (1988) have used Lippard's and other model compounds for their analyses of the Extended X-ray Absorption Fine Structure (EXAFS) of haemerythrin. EXAFS provides information about the immediate environment of metal atoms and requires no crystals, but since the fine structure arises from interference between waves scattered by neighbouring atoms, its interpretation requires knowledge of phases. If the environment of the metals is as complex as in haemerythrin, the phases can be solved only by reference to analogous model structure. Zhang *et al.* show that the EXAFS of oxyhaemerythrin is similar to that of Lippard's μ-oxo bridged complex, and the EXAFS of deoxyhaemerythrin to that of his OH-bridged complex. Using the atomic coordinates of the models to phase their EXAFS, they derived Fe—Fe distances of 3·24 Å for oxy- and 3·57 Å for deoxy-haemerythrin. Such an expansion of the Fe—Fe distance would be sufficient to produce weak cooperativity, similar to that exhibited, for example, by cobalt haemoglobin, where the distance of the metal from the haem plane changes on oxygenation by only 0·33 Å, compared to 0·55–0·60 Å in iron haemoglobin, and the free energy of cooperativity is only between a half and a third of that of iron haemoglobin (Fermi *et al.* 1982; Imai *et al.* 1977). It will be interesting to discover the change in quaternary structure of haemerythrin and to find out why some species bind oxygen cooperatively and others not, especially since both kinds are octameric and exhibit similar electronic absorption, circular dichroism and resonance Raman spectra; moreover, the oxygen affinity of the R-state of the cooperative ones is similar to the overall oxygen affinity of the non-cooperative ones (Richardson *et al.* 1983). P. R. Evans has suggested to me that an allosteric effector necessary for cooperativity might have been lost in the studies of the latter; a search for this may solve the riddle.

5. GLYCOGEN PHOSPHORYLASE: CONTROL OF GLYCOLYSIS

Phosphorylase is the key enzyme in the mobilization of chemical energy from glycogen. It is a complex allosteric protein that is subject to activation and inhibition by many chemical stimuli, including those of two other enzymes: phosphorylase kinase which activates it by phosphorylation of one specific pair of serine residues (No. 14), and phosphorylase phosphatase which inhibits it by hydrolysis of the serine phosphate bonds. In muscle, phosphorylase kinase is activated normally by the same release of calcium ions from the sarcoplasmic reticulum that also stimulates contraction. When activated, phosphorylase catalyses the stepwise phosphorylysis of glycogen with release of glucose-1-phosphate. Under *in vivo* conditions it is a dimer of two identical subunits, each containing a single polypeptide chain of 842 amino acid residues to which a pyridoxal phosphate is attached by a Schiff base at lysine 680. Its unphosphorylated form, known as phosphorylase *b*, is inactive under *in vivo* conditions, but can be activated *in vitro*, weakly by inosinemonophosphate (IMP) and strongly by adenosinemonophosphate (AMP), when it reaches 80 percent of the activity of the phosphorylated form. This form, known as phosphorylase *a*, exhibits nearly maximal activity without AMP. Each of the two forms is subject to regulation by other effectors, as shown below. The catalytic function of phosphorylase and many of its reponses to its regulators are cooperative (Graves & Wang, 1972; Fletterick & Madsen 1980; Fletterick & Sprang, 1982; Krebs, 1986; Madsen, 1986; Johnson *et al.* 1988).

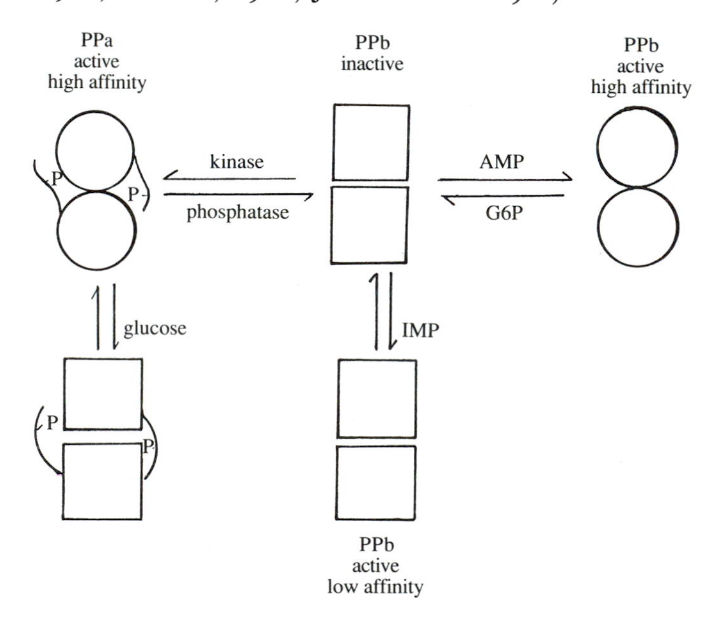

How does this cooperativity work? How do the effectors influence the allosteric equilibria? How can phosphorylation of a single pair of serine residues activate an enzyme of 1684 amino acid residues? Answers to these questions have come from X-ray analysis of two forms of the enzyme in the T-state: phosphorylase *b* partially activated by the weak effector IMP and phosphorylase *a* inhibited by glucose; and one structure in the R-state: phosphorylase *b* activated by ammonium sulfate.

Phosphorylase has a very complex structure. Each of its two subunits consists of two domains made up of a core of pleated β-sheets flanked by α-helices. The N-

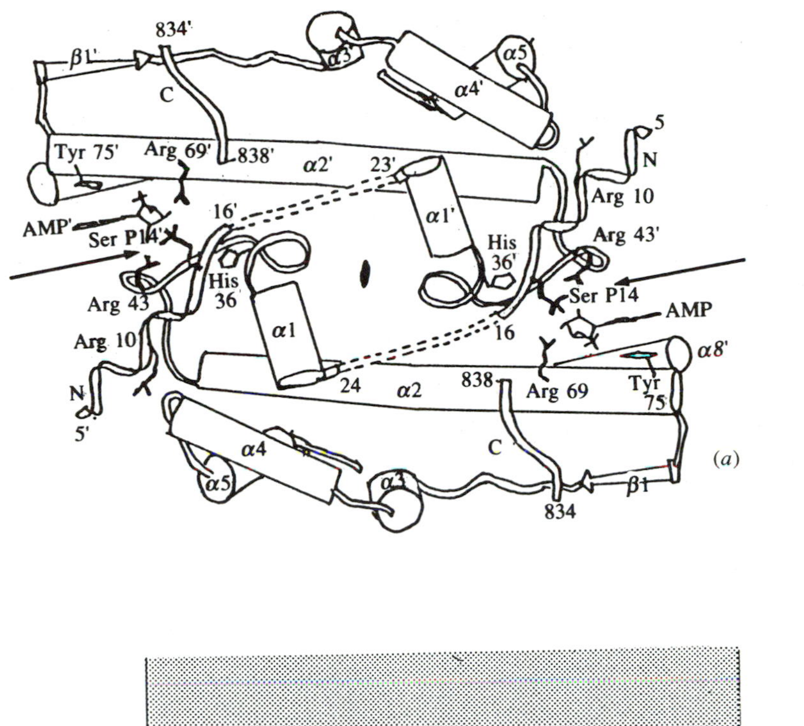

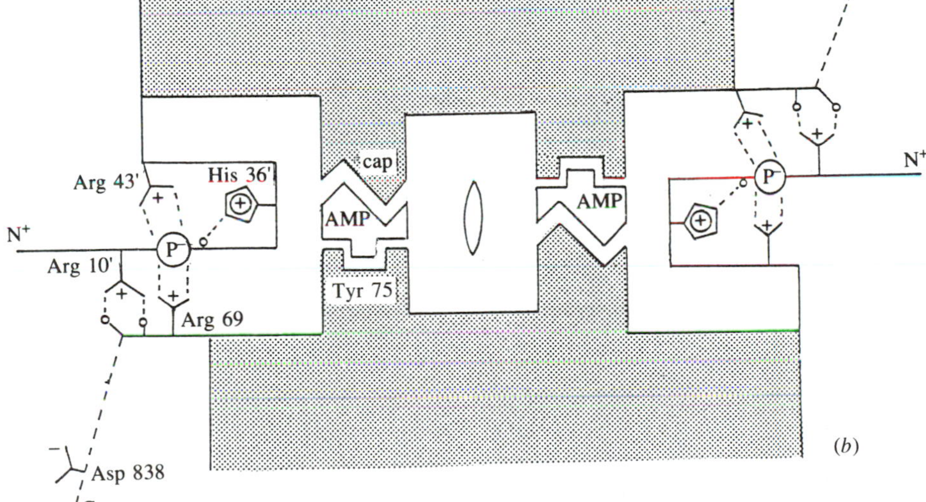

Fig. 26 (*a* & *b*). For legend see page 37.

terminal domain includes the subunit boundary, the serine phosphate, the activating AMP and inhibiting glucose-6-phosphate (G6P) binding site, the glycogen storage site and a small part of the catalytic site. The C-terminal domain complements the catalytic site and also contains the neighbouring site where the inhibitory nucleosides and purines bind. The catalytic site lies at the head of a 12–15 Å long tunnel (Fig. 25, Plate 1).

In the dimer the two subunits are joined end-to-end at a contact that is tenuous for so large a protein, making up no more than 7% of its surface area in

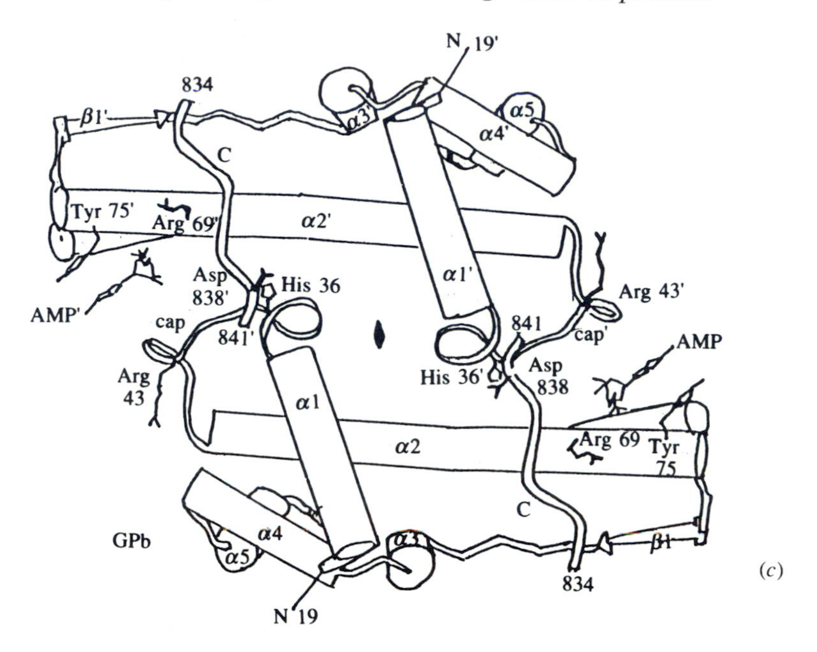

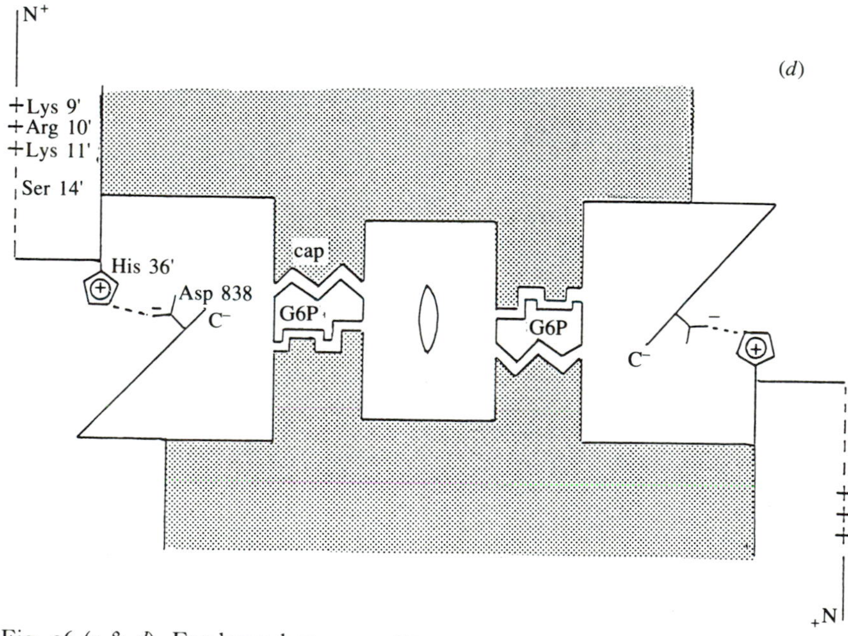

Fig. 26 (*c* & *d*). For legend see page 37.

phosphorylase *b* and 10% in *a*. As in haemoglobin, the twofold symmetry axis passes through a water-filled channel, but in phosphorylase this channel is flanked by two large grottoes, capable of holding about 150 water molecules. Unlike haemoglobin, the channel contains no binding site for effectors. A view of the dimer perpendicular to the symmetry axis shows that its two sides are very differently constructed. One side is convex with a radius of curvature matching

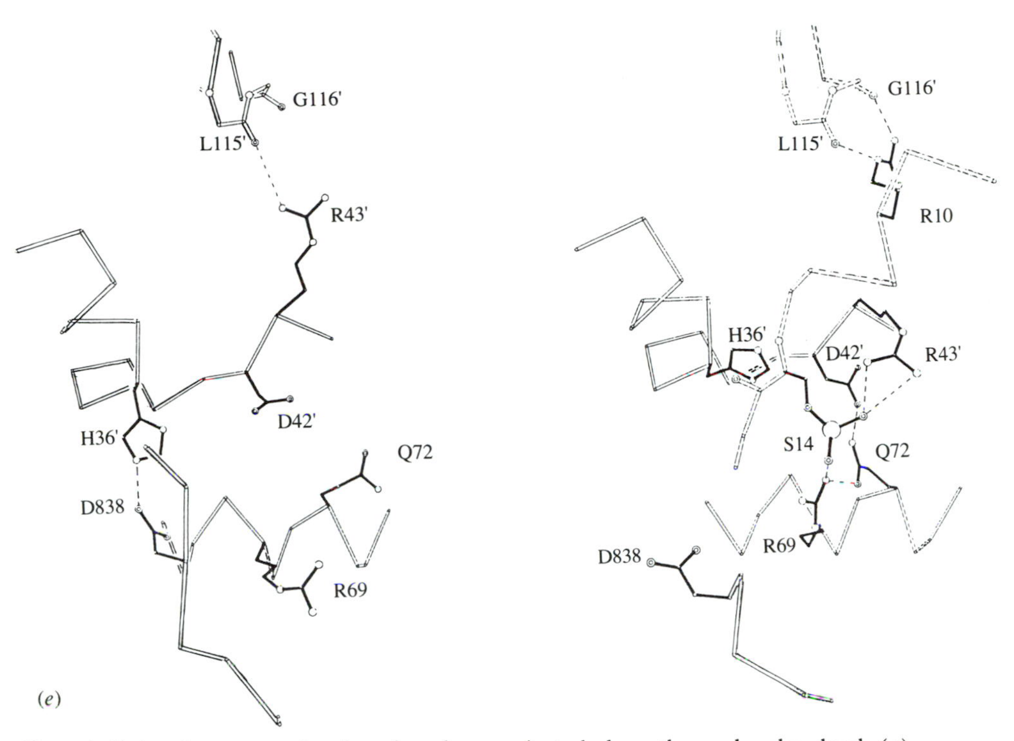

Fig. 26. Subunit contacts in phosphorylase projected along the molecular dyad. (*a*) Phosphorylase *a* showing the N-terminal helices, marked *N*, with the serine phosphates 14, marked by arrows, coordinated to two arginines, one from each subunit and AMP bound firmly between helix α8 and the cap. The C-terminal peptide, marked C, is disordered. (*b*) Same view in diagrammatic form. (*c*) Phosphorylase *b*. The N-terminal residues are disordered and the C-terminal ones are ordered, with Asp 838 hydrogen bonded to His 36 of the opposite subunit. AMP is bound more loosely (L. N. Johnson, private communication). (*d*) Diagrammatic view, with G6P bound at the effector site. The plus signs stand for Lys 9, Arg 10 and Lys 11. (*e*) Enlarged views of subunit contacts in phosphorylase *b* (left) and *a* (right), showing details of some of these interactions. (From Sprang *et al.* 1988.)

that of the glycogen particle (175 Å). It contains the entrance to the catalytic tunnel and the glycogen storage site, identified by its binding of maltoheptaose, a 7-residue oligomer of α-D-glucose. The side that faces away from the glycogen particle contains the regulatory phosphorylation sites and the overlapping AMP and G6P binding sites. Most of the binding sites for substrates and effectors are widely separated. A distance of 30 Å separates the catalytic site from the one that binds maltoheptaose (the glycogen storage site), which suggests that the enzyme chews away at a piece of polysaccharide chain that is far removed from the piece that attaches the glycogen particle to the enzyme. The closest distance, 15 Å, links the serine phosphate to the AMP binding site; both are over 30 Å from the nearest catalytic site, and the two catalytic sites are over 60 Å from each other. Yet binding of ligands to any of these sites can be shown to affect all the others. Interpretation of this complex behaviour presents a great challenge to X-ray crystallography.

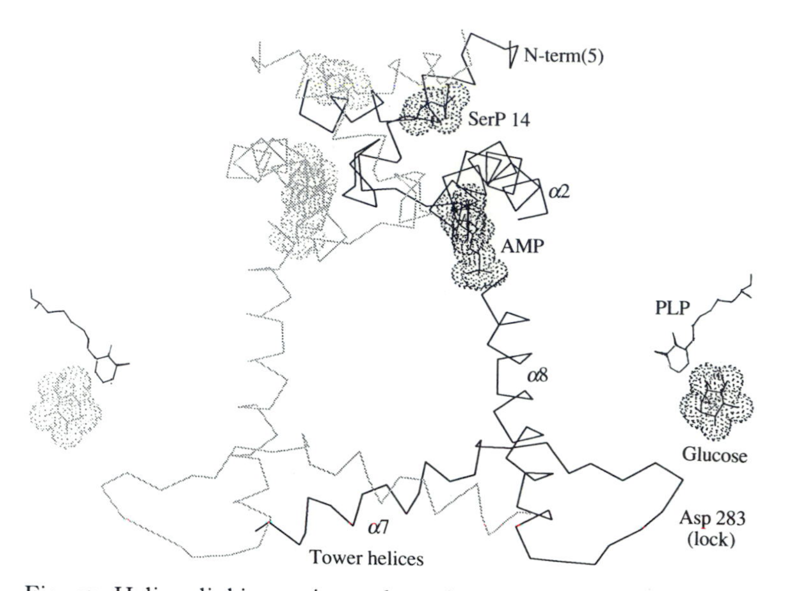

Fig. 27. Helices linking active and regulator sites of phosphorylase in the T-structure seen perpendicular to the molecular dyad (Courtesy Dr R. J. Fletterick).

As a first step, X-ray analysis has revealed the changes that phosphorylation of serine 14 induces at the subunit boundary in going from the weakly activated *b* to the inhibited *a* structure. The most important of these changes consists in the burial and ordering of the amino-terminal 16 residues and the exposure and disorder of the carboxy-terminal 5 residues in phosphorylase *a*, and the reversal of these features in phosphorylase *b* (Fig. 26). These movements are accompanied by changes in hydrogen bonding. In *b*, Asp 838 is tied down by a salt bridge to His 36 of the opposite subunit. On transition to *a* that salt bridge is broken, and the histidine rotates about the α-β bond to form a hydrogen bond with the carbonyl oxygen of phosphoserine 14 of the same subunit. The phosphate also forms salt bridges with two arginines, one from the same and the other from the opposite subunit. In the absence of the neutralizing serine phosphate, these arginines contribute to a cluster of positive charges that expel the positively charged N-terminal peptide from its binding site that spans the two protein subunits. Thus the dominant interactions responsible for the allosteric transition are electrostatic.

Fig. 28. The allosteric transition as it affects helices $\alpha7$ (the towers) and $\alpha8$, the loop of chain connecting them (the lock) and the pyridoxal phosphates; (*a*) in the T and (*b*) in the R structure, in projection normal to the molecular dyad; (*c*) in the T and (*d*) in the R-structure projected along the molecular dyad. Note the hydrogen bonds between the asparagines linking the tower helices. Arginines 310 form hydrogen bonds with the phosphates of AMP. The pyridoxal phosphates are shown attached to lysines 680. This figure has been turned by 90° with respect to Fig. 27 in order to make the labels clear. The change in the angle between the symmetry-related helices $\alpha7$ shown in (*a*) and (*b*) results from the rotation of the two subunits about axes normal to the twofold axis as indicated in Fig. 1, p. 3, and from a change in tilt of each helix relative to its own subunit. (D. Barford & L. N. Johnson, 1989.)

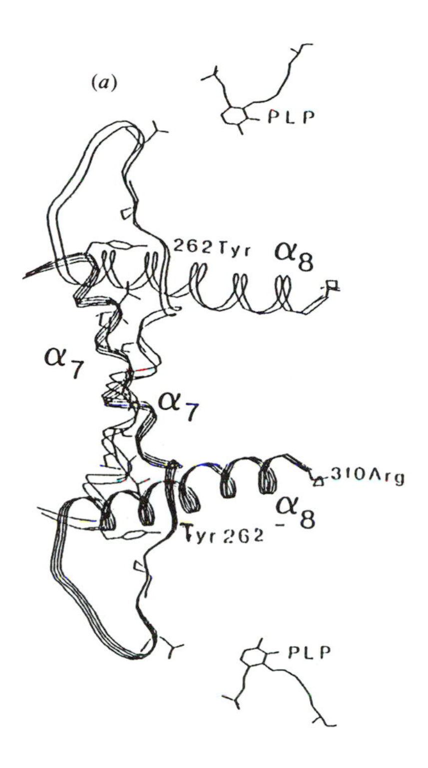

T

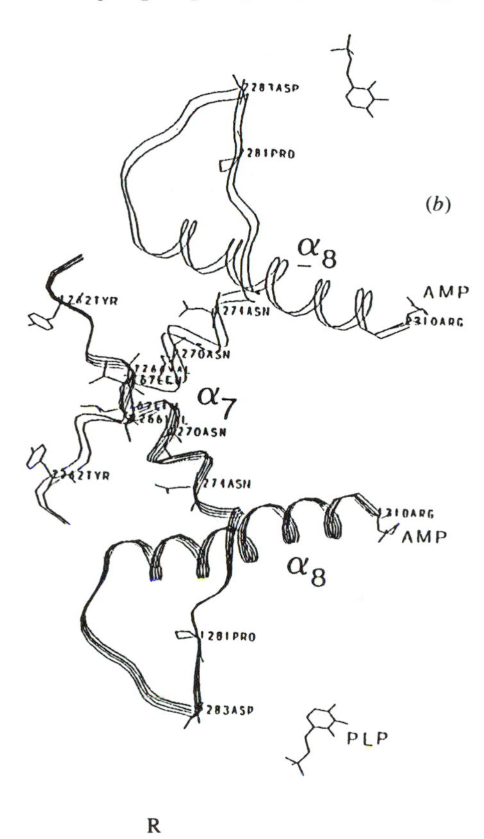

R

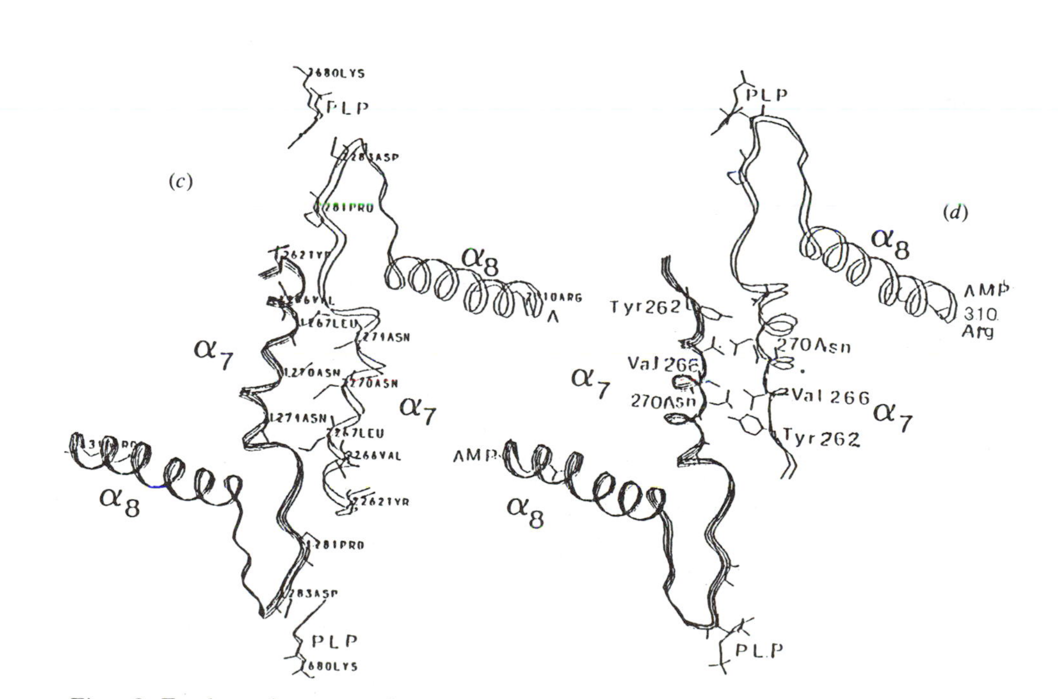

Fig. 28. For legend see opposite.

Other changes in salt bridges and non-polar contacts between subunits follow in train, and these are transmitted to the AMP- and G6P-binding sites that lie at the subunit boundary only 15 Å from phosphoserine 14. AMP is wedged between an α-helix from one subunit and a non-helical loop from the opposite subunit, referred to as the cap: that loop is separated from phosphoserine 14 by a short helix. Comparison of the two structures shows how dephosphorylation of serine 14 and transition to phosphorylase *b* weakens both electrostatic and van der Waals interactions with the activator AMP, thus increasing its dissociation constant from the enzyme 100-fold. The same changes strengthen binding of the inhibitor G6P (Sprang *et al.* 1988; Perutz, 1988). All these changes take place in the T-structure.

D. Barford and L. N. Johnson have now solved the long-sought structure of the active R-form of phosphorylase. It emerged from an X-ray analysis of crystals of phosphorylase *b* grown from ammonium sulphate solution; the sulphate ion appears to have worked as an activator in place of phosphate at the active site and at the phosphorylation site at serine 14. In these crystals the enzyme is tetrameric, a form of phosphorylase *b* also found in solution, and the glycogen storage site is buried in a subunit contact. *In vivo* attachment of the enzyme to glycogen particles causes it to dissociate into dimers.

The allosteric transition consists of rotations of each of the two monomers by 5° about axes pointing in opposite directions normal to the molecular dyad, as shown in Fig. 1, p. 3. The transition affects the helices $\alpha7$. They inter-digitate and form a bridge between the catalytic sites of neighbouring subunits; the authors call them tower helices, because each helix protrudes from its own subunit and penetrates deeply into the neighbouring one (Fig. 27). The angle between the two helices changes from +20° in the T structure to −80° in the R structure (Fig. 28). Each catalytic site is flanked by six loops of chain; some from the N-terminal and some from the C-terminal domain (Fig. 29). One of these loops links helix $\alpha7$ to $\alpha8$; it carries aspartate 283, asparagine 284 and phenylalanine 285 at its tip, which lock access to the catalytic site when the enzyme is inhibited. Helix $\alpha8$ reaches from the catalytic site to the AMP binding site of the same subunit. At its C-terminus Arg 309 and 310 form hydrogen bonds with the phosphate of AMP. Finally helices $\alpha2$ link each of the AMP binding sites to the phosphoserine sites on the opposite subunit.

Fig. 28 shows two orthogonal views of the disposition of the helices $\alpha7$ and $\alpha8$ in the T and R structures. Here is Barford and Johnson's own description of the allosteric transition:

> In the T-state the tower helices pack antiparallel with a cluster of hydrogen bonds between asparagines 270 and 274 and their counterparts in the other subunit. Tyrosine 262 is in van der Waals contact with proline 281 at the start of the 280s loop (the lock) of the other subunit. This loop blocks access to the catalytic site in the T-state; aspartate 283 is in indirect contact (through two water molecules) with the 5′-phosphate of the essential cofactor pyridoxal phosphate. In the R-state the tower helices change their translation and angle of tilt with respect to each other. The asparagine/asparagine contacts are broken; instead, valine 266 and isoleucine 267 are now opposite their counterparts in the other subunit. There are no longer any contacts between the top of

the tower (e.g. tyrosine 262) and the 280s loop of the other subunit. The 280s loop is disordered and no longer blocks access to the catalytic site. These movements enable ionic residues at the catalytic site to adopt their correct orientation to promote substrate binding and catalysis, and they provide a mechanism for transmission of homotropic allosteric effects.

Heterotropic ones may be transmitted to the regulatory sites by the tower helices and by changes at the subunit contacts of the kind shown in Fig. 26 (Barford & Johnson, unpublished results).

Goldsmith *et al.* (1989) have solved the structure of a crystal of phosphorylase *a* soaked in a solution of orthophosphate and maltopentaose. Phosphate ions were bound at the catalytic and nucleotide activator sites, while maltopentaose was bound at the glycogen storage rather than the catalytic site. The binding of these molecules induced marked changes in the enzyme's structure. At the active site the phosphate ion unlocked the 'gate', displaced Asp 283 and formed hydrogen bonds with imino groups at the end of the helix containing residues 133–149. (The imino groups of glycine 135 and leucine 136 are seen on the right of the phosphates in Fig. 29.) The binding of the phosphates and of the oligosaccharide caused the C-terminal catalytic domain to turn by 1° and shift by 0·5 Å away from the N-terminal regulatory domain; the tower helices moved closer together. Their movement was much smaller than, and different from, the one seen by Barford and Johnson in the R-structure of phosphorylase *b*. Goldsmith *et al.* suggest that their structure may be an intermediate between the inactive T and the fully active R-structure.

First glimpses of the enzyme's role in catalysis came from X-ray studies that showed how the substrate phosphate, the reaction product glucose-1-phosphate and the inhibitory cyclic glucose-1,2-biphosphate bind to phosphorylase *b* and *a*. The most telling clues were obtained from synchrotron radiation studies of crystals of phosphorylase *b* activated by AMP and soaked in solutions of the sugar heptenitol and inorganic phosphate. The very intense X-ray beam from the synchrotron storage ring allowed the investigators to take 'snapshots' of the diffraction pattern at successive stages of the reaction, and to analyse the changes that accompanied the gradual accumulation of the product heptulose-2-phosphate that remains bound in the active site, as shown in Fig. 29 (Hajdu *et al.* 1987). The conformations of the protein and coenzyme seen in this figure are similar to those found more recently in the R-structure, where the substrate phosphate is replaced by a sulphate ion and the sugar is absent.

In phosphorylase it is not the pyridoxal moiety but the phosphate moiety of pyridoxal-phosphate that activates the substrate. It assumes the monoanionic form in the inactive T-state and the dianionic form in the active R-state (Helmreich & Klein, 1980). The binding site for the substrate inorganic phosphate lies right next to the pyridoxal phosphate, and both are hydrogen-bonded to basic groups of the protein. The binding of heptulose-2-phosphate causes arginine 569 to move from a buried position to another that is close to the coenzyme and product phosphates, displacing aspartate 283, the 'lock', in the

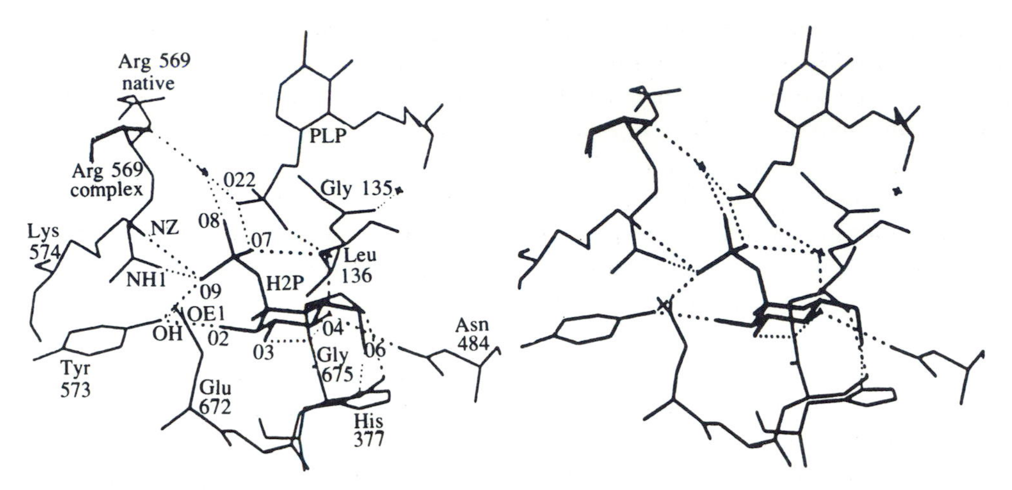

Fig. 29. Active site showing pyridoxal phosphate (PLP) and heptulose-2-phosphate (H2P) hydrogen-bonded to the enzyme (Hajdu *et al.* 1987).

process, thus substituting a buried negative, repulsive charge by a positive one that attracts the phosphates. The movements of lysine 574 and arginine 569 are also seen in the R-structure and may stabilize the extra ionization of the pyridoxal phosphate that is essential for catalysis. Glutamate 672 has been shown to be essential as a proton acceptor (Helmreich, private communication).

Chemical and structural data suggest that catalysis *in vivo* may involve the steps shown below. First, a proton may be transferred from the coenzyme phosphate to the substrate phosphate. The latter may then act as a general acid, protonating the α-(1,4) glycosidic bond that links the terminal glucose to the glycogen chain. Cleavage of that bond would lead to the formation of an oxocarbonium ion on the free sugar which would be stabilized by the newly formed orthophosphate dianion. Finally, the orthophosphate may transfer its proton to the coenzyme phosphate and simultaneously mount a nucleophilic attack on the oxocarbonium ion, forming glucose-1-phosphate. In this mechanism the coenzyme phosphate plays the dual part of a general base catalyst (Palm *et al.* 1989; Barford *et al.* 1988).

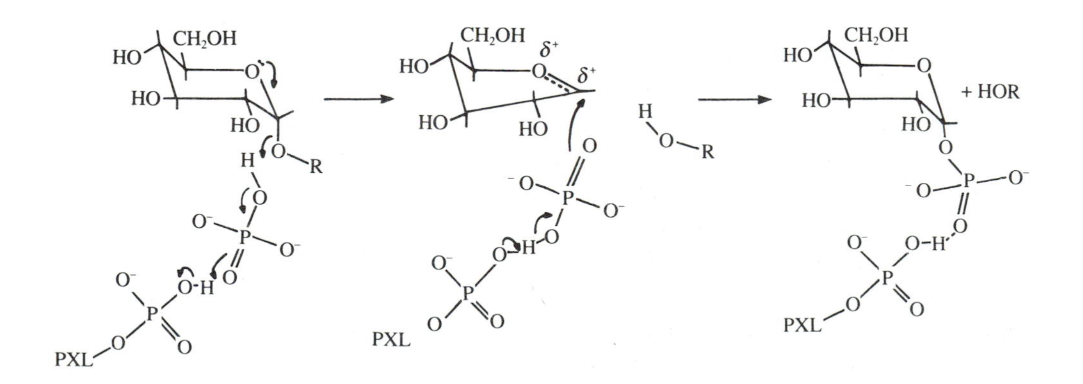

Phosphorylase is a more complex allosteric system than either haemoglobin or phosphofructokinase, because both *a* and *b* can take up at least two alternative quaternary structures. Sprang and Fletterick suggest that some of the crystal structures analysed by X-rays may be chimeric in the sense that part of each subunit approaches the tertiary R-structure and another part the tertiary T-structure. This may apply to the heptulose-2-phosphate complex of *b*, where the catalytic site has the active and the subunit boundary the inactive form, or to the glucose complex of *a*, where the catalytic site has the inactive structure, while the subunit boundary has a structure close to fully active R.

In summary, the allosteric transition from the T- to the R-structure can be set in train by phosphate or sulphate ions buried at the phosphorylation and catalytic sites. The transition consists of a rotation of one subunit relative to the other by 10° about an axis at the subunit boundary that is normal to the twofold symmetry axis. Allosteric effects are transmitted from the subunit boundary to the active site by the tower helices $\alpha 7$ which tilt and slide relative to each other. These helices terminate in a polypeptide loop that rigidly locks access to the catalytic site in the T-structure. In the R-structure that loop becomes disordered. Its displacement allows substrates as well as additional cationic side chains to move into the catalytic site. They convert the coenzyme phosphate from the mono to the dianionic form that is necessary for catalysis. Helix $\alpha 8$ links the catalytic to the AMP-binding site suggestive of tertiary interactions between them, but so far these have not been observed.

6. PHOSPHOFRUCTOKINASE: FURTHER CONTROL OF GLYCOLYSIS

6.1 *The classic experiments*

Phosphofructokinases (PFK's) control the rate of glycolysis in the cell. The bacterial enzymes consist of four identical subunits of molecular weight of about 35000. They catalyse the phosphorylation of fructose-6-phosphate (F6P) to the 1,6-biphosphate ($F1,6P_2$) by transfer of the γ-phosphate of ATP.

Blangy *et al.*'s (1968) paper on the kinetics of the allosteric interactions of PFK from *E. coli* is one of the classics of molecular biology. Here is a summary of their chief observations. In 0·1 mM ATP + 1 mM $MgCl_2$ the rate of the reaction v/V_{max} was highly cooperative with respect to [F6P], with a Hill's coefficient of 3·8, defined by the slope of (v/V_{max}) versus log [F6P]. On the other hand, the rate was non-cooperative with respect to [Mg-ATP] at all concentrations of F6P, and the concentration of F6P needed to reach $V_{max}/2$ was independent of [Mg-ATP] over a wide range of [Mg-ATP]. The initial velocity rose and Hill's coefficient fell with rising concentrations of the activators ADP and GDP. At very high F6P concentrations ADP acted as a competitive inhibitor of ATP, but GDP worked as an activator at all F6P concentrations. When [F6P] was low, the initial velocity of the enzymic reaction was cooperative with respect to [GDP]. At high [F6P] the initial velocity was high, independent of [GDP]. Phosphoenolpyruvate (PEP) acted as an inhibitor, reducing both the initial velocity and the cooperativity of the enzymic reaction. Neither GDP nor PEP altered the maximum rate of that

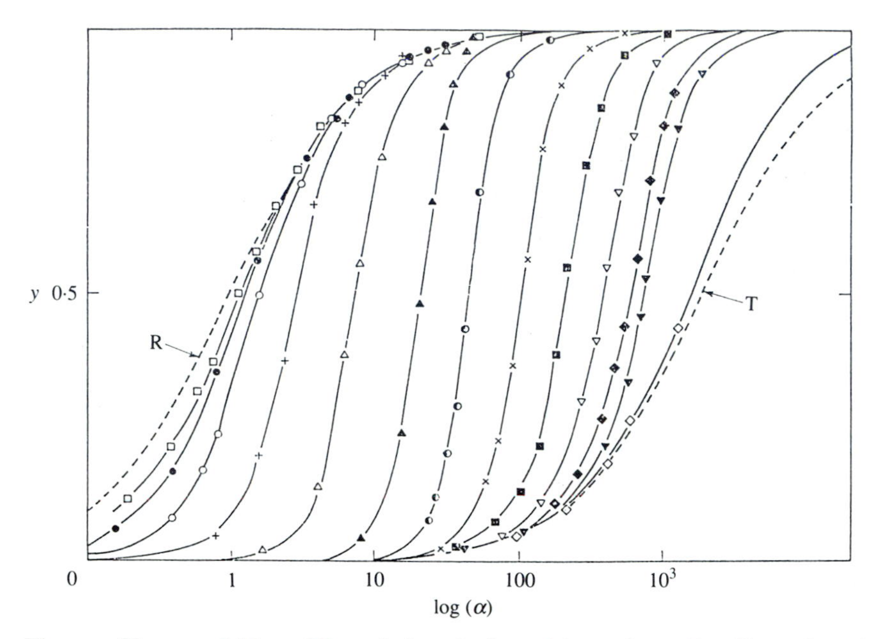

Fig. 30. Change of $Y = v/V_{max}$ of phosphofructokinase from *E. coli* as a function of log α, where $\alpha = [F6P]/1{\cdot}25 \times 10^{-5}$. The experimental points were obtained at different concentrations of ADP or phosphoenolpyruvate. The theoretical curves were calculated from the equations of Monod *et al.* (1965), taking $n = 4$ and $c = K_R/K_T = 5 \times 10^{-4}$ (from Blangy *et al.* 1968).

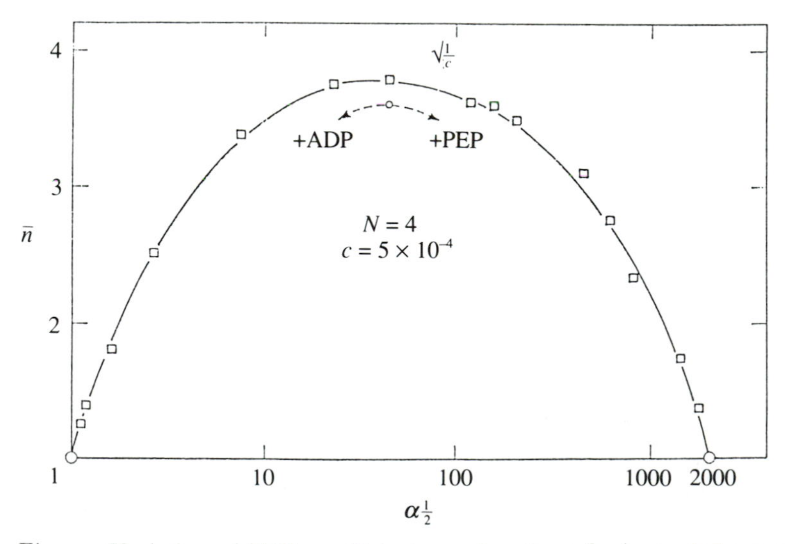

Fig. 31. Variation of Hill's coefficient as a function of $\alpha/2$ needed to reach half-maximum velocity. $\alpha_{\frac{1}{2}} = [F6P]_{\frac{1}{2}}/1{\cdot}25 \times 10^{-5}$. Each point corresponds to one of the curves in Fig. 30. (From Blangy *et al.* 1968.)

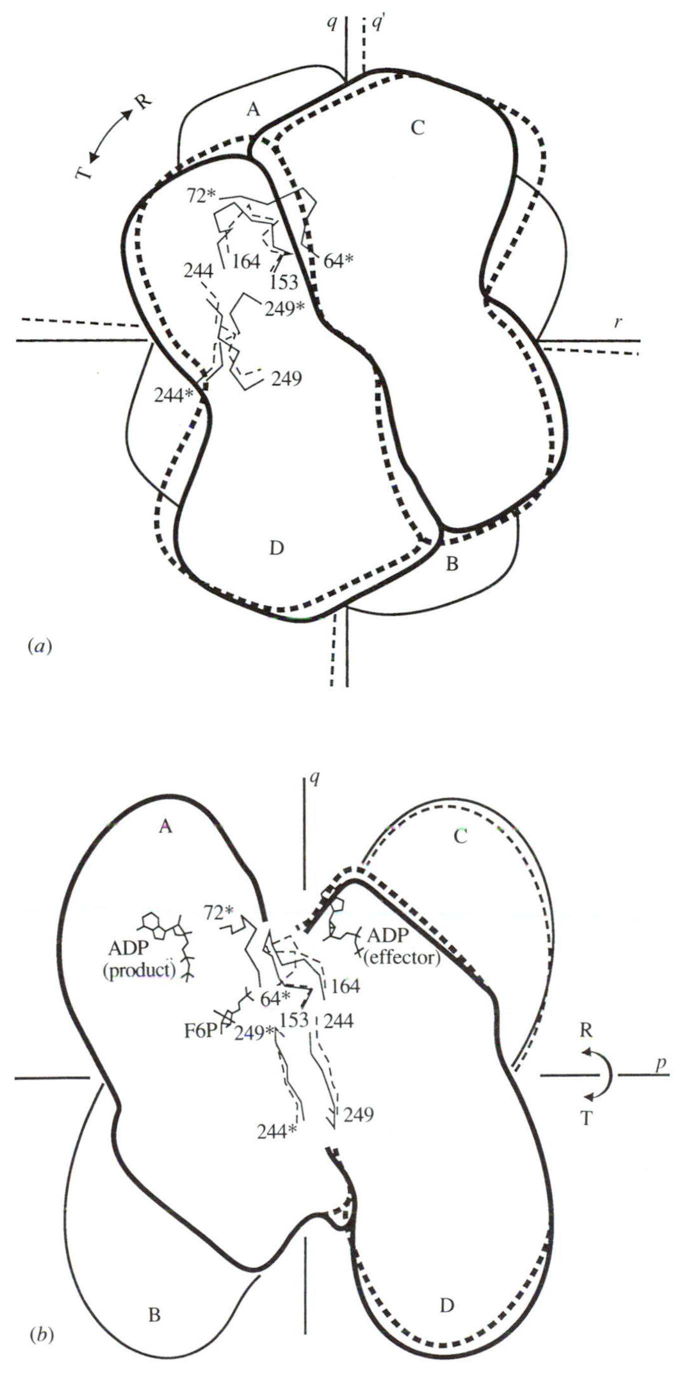

Fig. 32. Outlines of phosphofructokinase with its four subunits linked rigidly in pairs, projected along the *p*-axis in (*a*) and the *r*-axis in (*b*). On transition from the T-structure (full lines) to the R-structure (broken lines) one pair turns relative to the other by 7° about the *p*-axis. The projections trace the α-carbon positions of residues 72*–64* in subunit A, facing residues 153–164 in subunit D, and of residues 244*–249* in A facing 244–249 in D. (From Schirmer & Evans, 1989.)

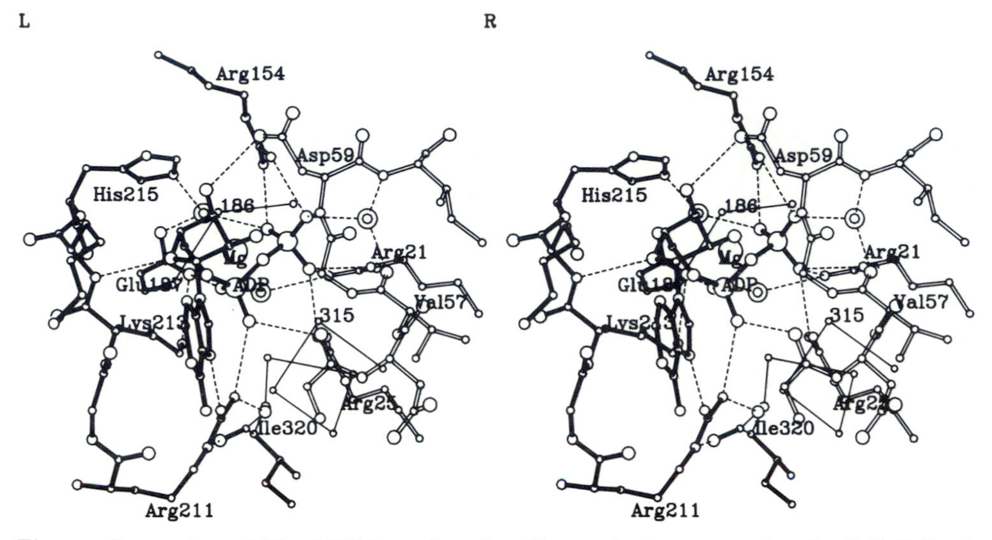

Fig. 33. Stereoview of Mg-ADP bound to the effector site between subunits C (full lines) and D (broken lines) in the R-structure. The carboxylate of Glu 187 is bound to the Mg^{2+} ion and to Lys 213. (From Schirmer & Evans 1989.)

reaction, but both affected the substrate concentration needed to reach that rate. Fig. 30 illustrates the dependence of v/V_{max} on the concentration of F6P at various effector concentrations. The family of curves is similar to the oxygen equilibrium curves of haemoglobin at various effector concentrations shown in Fig. 5. Cooperativity is lost at high concentrations of either the activator or the repressor. V_{max} is independent of the effectors; they only affect the concentration of F6P needed to attain V_{max}. Fig. 31 shows the bell-shaped curve obtained when Hill's coefficient n is plotted against the concentration of F6P needed to reach $\frac{1}{2}V_{max}$; it is similar to the plot of L against n for haemoglobin in Baldwin (1975). Mathematical analysis led Blangy, Buc and Monod to conclude that

(i) The enzyme is made of four equivalent protomers, each bearing a single site for each ligand.

(ii) Two distinct conformational states in equilibrium are accessible to the protomers.

(iii) The transitions in each molecule are fully concerted.

(iv) The two states have the same affinity for ATP, but differ with respect to their affinity for F6P, XDP and PEP. These ligands bind independently to the two conformational states.'

The recently determined structures of the two states of bacterial PFK have brilliantly confirmed these conclusions, bar the last sentence, and have accounted for almost every detail of the enzyme's complex behaviour.

6.2 *Three-dimensional structures*

P. R. Evans and his collaborators have solved the structures of the PFK from *B. stearothermophilus* and of the major PFK isozyme from *E. coli* (Evans & Hudson,

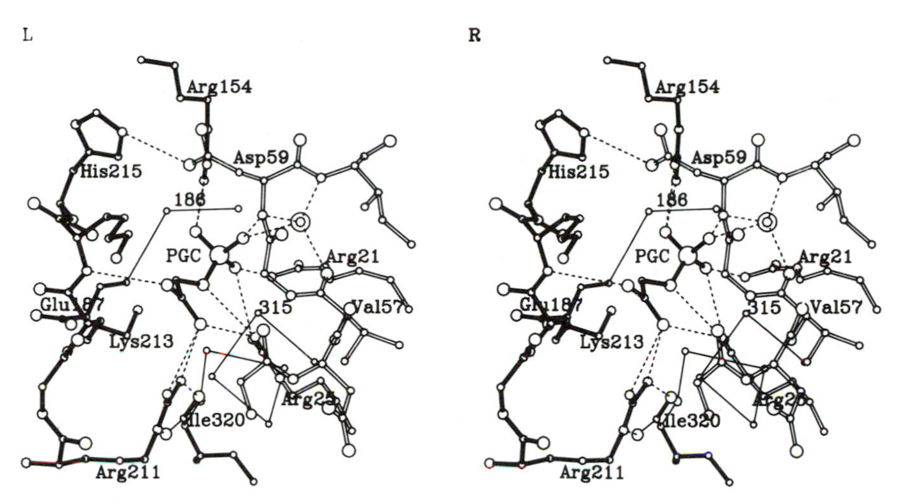

Fig. 34. Stereoview of phosphoglycolate (PGC) bound to the same effector site as in Fig. 33 in the T-structure. The carboxylate of Glu 187 has turned away from the effector and from Lys 213. (From Schirmer & Evans, 1989.)

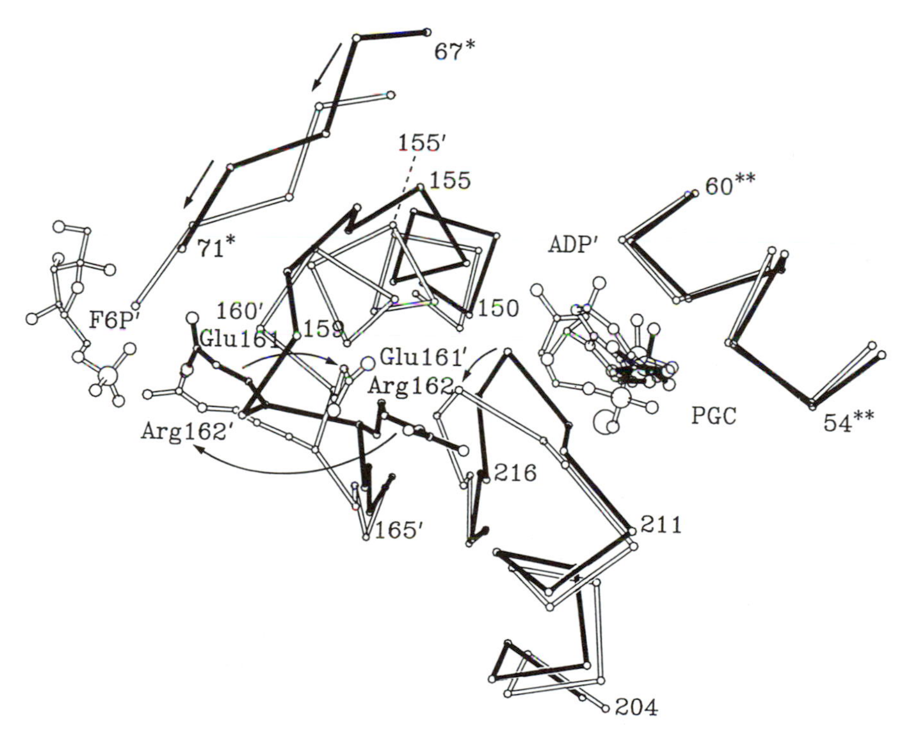

Fig. 35. Key changes between the effector and catalytic sites of phosphofructokinase in going from the T-structure (thick lines and plain labels) to the R-structure (thin lines and primed labels). Residues without stars are from subunit D, with one star from A, with two stars from C. Note the replacement of Arg 162' facing the catalytic site in the R-structure by Glu 161 in the T-structure. (From Schirmer & Evans, 1989.)

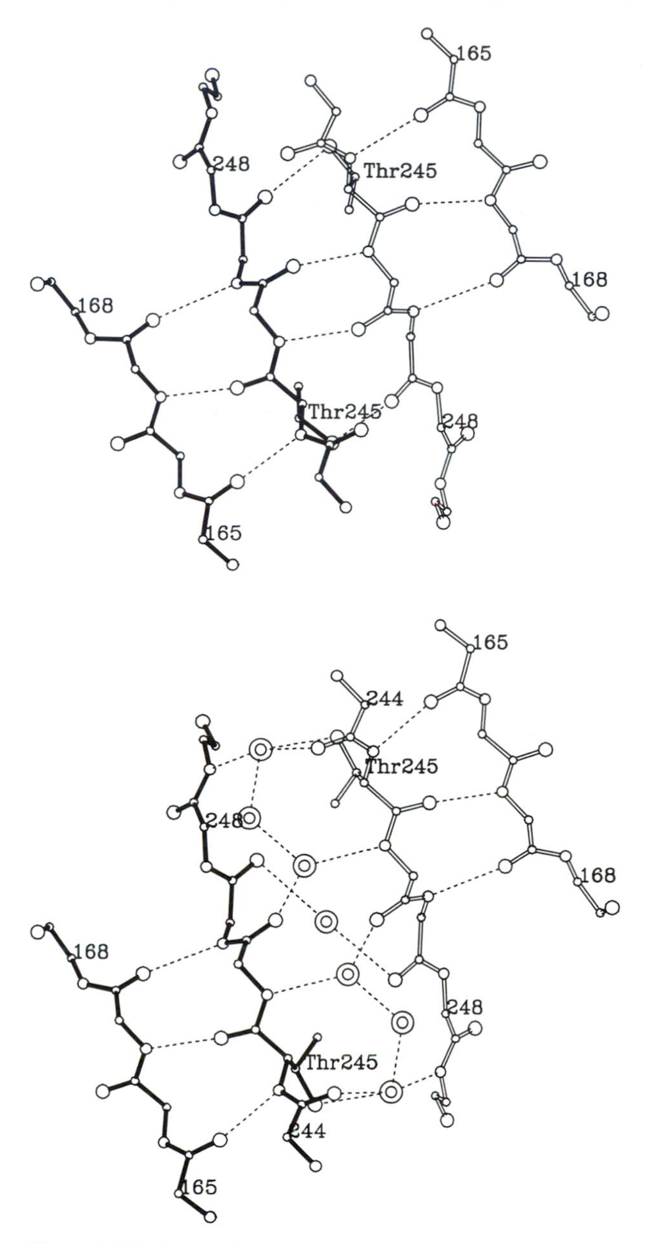

Fig. 36. Hydrogen bonds between subunits A and D as seen in Fig. 32*(b)*, p. 45; in *(a)* the T- and *(b)* the R-structures. (From Schirmer & Evans, 1989.)

1979; Evans *et al.* 1986; Kotlarz & Buc, 1982; Shirakihara & Evans, 1988; Schirmer & Evans, 1989). They and A. R. Fersht have also explored the reaction mechanism by directed mutagenesis of the *E. coli* enzyme (Hellinga & Evans, 1987; Lau & Fersht, 1987; Lau *et al.* 1987).

The four subunits of PFK occupy the corners of a squashed tetrahedron and are related by three mutually perpendicular axes of symmetry (Fig. 32). They are elongated and divided into two domains, a large and a small one, each one consisting of a core of pleated sheet flanked by α-helices, characteristic of nucleotide binding enzymes. Cooperativity arises by changes in tertiary and quaternary structure that affect the catalytic and effector sites. The four catalytic sites span the boundaries between neighbouring domains and between the subunits that are related by the twofold symmetry axis parallel to *r*. The effector sites span the boundaries between subunits that are related by the twofold axis parallel to *p*. The binding sites for activators and inhibitors overlap, so that the enzyme contains only four effector sites, rather than the eight envisaged by Blangy *et al.* (1968).

In the active (R) structure, the activator Mg-ADP is bound by hydrogen bonds to polar groups from subunit D and from the p-axis-related subunit C (Fig. 33). Even though the regulator sites lies at the interface between the *p*-axis-related pairs of subunits, replacement of the activator Mg-ADP by the inhibitors phosphoenolpyruvate or phosphoglycolate leaves that interface unaltered (Fig. 34). Instead it induces changes in tertiary structure that make themselves felt at the boundary between the subunits related by the *r*-axis, causing causing one pair of subunits to rotate relative to the other pair about the symmetry axis *p* by 7° (Fig. 35). In the active R-structure, parallel β-strands of neighbouring subunits are linked at that boundary by hydrogen-bonded water molecules; in the inactive T-structure these are expelled and the β-strands are linked by direct hydrogen bonds (Fig. 36). The inclusion or exclusion of water is an all-or-none effect that acts like a two-way switch. The 7° rotation directly affects the catalytic sites which lie at the same subunit boundary. At the catalytic site in subunit *A*, ATP binds in a pocket of the larger domain, while F6P occupies a pocket in the neighbouring smaller domain; in the R-structure the 6-phosphate forms hydrogen bonds with an arginine and a histidine of that small domain, and also with arginines 162 and 243 of the neighbouring subunit *D* (Fig. 37). Arginine 162 lies at the end of a helical turn which is unwound in the transition to the T-structure. As a result, glutamate 161 takes the place of arginine 162 and forms a hydrogen bond with arginine 243, thus neutralizing its charge; arginine 162 now points away from the catalytic site, so that the doubly-negative 6-phosphate anion loses both its cationic ligands and is repelled by glutamate 161. Another cation is lost to ATP: in the R-structure arginine 72 forms hydrogen bonds with the phosphate oxygens of ATP; in the T-structure it turns round and forms hydrogen bonds with glutamate 241 of subunit D instead. Finally, stereochemical changes in the F6P-binding domain make the substrate a misfit there. This explains the ratio of 5×10^{-4} between the dissociation constants of F6P from the R and T structures calculated by Blangy *et al.* (1968). The structure of the active site also explains why the kinetics of the reaction are cooperative with respect to [F6P] and not with respect to [ATP]: it shows that only the binding of F6P affects the subunit boundaries; ATP, being bound within a subunit domain, does not make itself felt there, and therefore does not affect the allosteric equilibrium.

Berger & Evans (1989) used directed mutagenesis to replace each of the arginines 162, 243 and 72 separately by serines and measured the reaction kinetics of the mutant enzymes in both directions (Table 1). In the forward reaction in 1

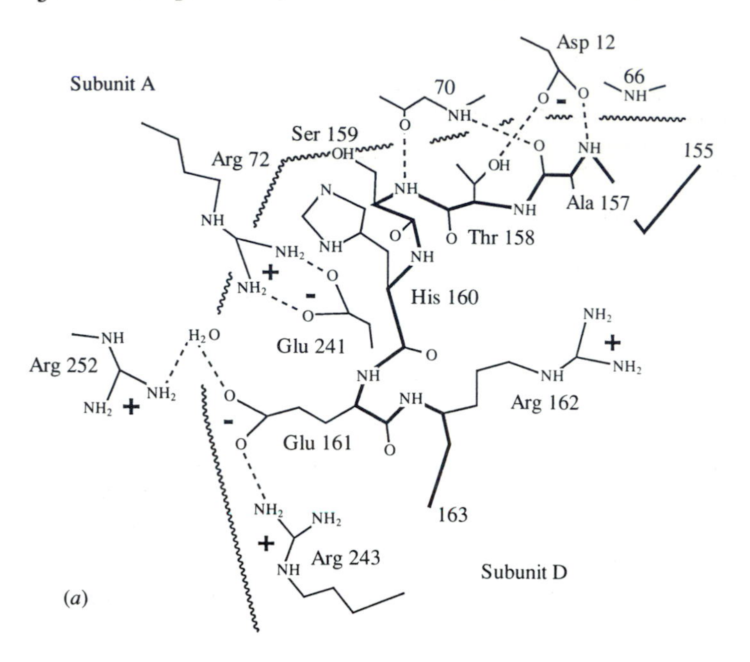

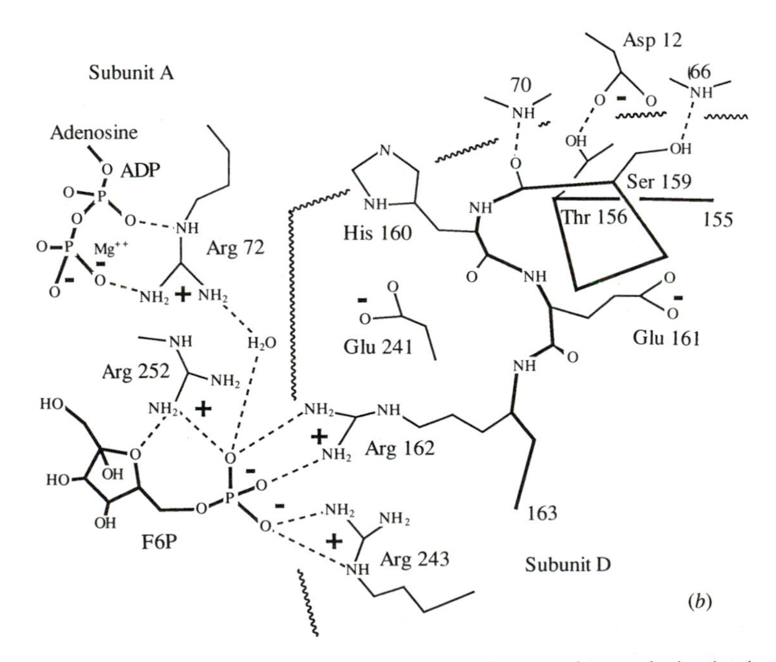

Fig. 37. Contact between subunits A and D near the catalytic site in (*a*) the T- and (*b*) the R-structures. In R, residues 156–159 make a helical turn; arginines 72, 162 and 243 bind the substrates. In T the helical turn is unwound, and the arginines are tied down by glutamates 161 and 241. Note the inversion of the positions of glutamate 161 and arginine 162. (From Schirmer & Evans, 1989.)

Table 1. Steady-state kinetic parameters of wild-type and mutant PFK's

	Forward reaction					Reverse reaction			
	k_{cat} (s^{-1})	$K_{m[F6P]}$ (μM)	$K_{m[ATP]}$ (μM)	$S_{\frac{1}{2}}$ (μM)	n_H	k_{cat} (s^{-1})	$K_{m[ADP]}$ (μM)	$S_{\frac{1}{2}[F1,6P]}$ (mM)	n_H
w.t.	134	30	63	540	4	22	52	1·9	1
RS162	95	4950	42	3700	2·1	5	124	28·1	2·0
RS243	186	1600	26	16000	2·7	8·9	41	13·0	2·0
RS72	4	96	75	700	2·2	1·5	56	39·5	1

mM GDP, the replacements raised K_M(F6P) 165-, 53- and 3-fold compared to the wild type. Without GDP they lowered Hill's coefficient and raised $S_{\frac{1}{2}}$, the concentration of substrate required to reach V_{max}. All these effects can be accounted for by a rise in L, due to the loss of hydrogen bonds in the R-structure. The reverse reaction of F1,6P_2 with ADP is non-cooperative in the wild type. In the mutants, K_M(F1,6P_2) is raised 15-, 7- and 21-fold, and in two of them the reaction becomes cooperative. This behaviour can also be accounted for by a rise in L. In the wild type ADP works as a strong activator, and makes the reverse reaction proceed in the R-state throughout, but it fails to do so in two of the mutants which evidently are in the T-state initially and are converted to R only as the reaction proceeds.

The construction of the enzyme makes the crucial tertiary changes at the catalytic site dependent on the rotation between the two dimers. If the dimers turned from T to R without unfolding the helical loop that carries arginine 162, then that loop would clash with the neighbouring dimer. Conversely, if the dimers turned from R to T and left the loop unfolded, then a large gap would remain (Fig. 38). This construction ensures the concerted nature of the allosteric transitions.

The binding of effectors does not rotate the dimers directly, but sets in train a set of tertiary changes that make themselves felt at the boundary between the two dimers, rather as the binding of haem-ligands makes itself felt at the contact between the α_1 and β_2 subunits. It moves a loop of chain centred on residues 212–216 that leans against another loop centred on arginine 162 and that in turn pulls the neighbouring dimer. The effector binding site is similar in R and T, except for glutamate 187, which binds to Mg^{2+} in R and turns away from the effector in T, forcing leucine 205 to make way for it (Figs. 33–35).

6.3 *Catalytic and regulatory mechanism*

Information about the reaction mechanism comes from a combination of chemical studies, protein engineering and X-ray analysis. In theory, phosphate could be transferred from ATP to substrate either directly or by being taken up first by an acceptor group of the enzyme and then transferred to the substrate. The two alternative mechanisms can be distinguished by the use of phosphate with isotopically labelled oxygens: ^{16}O, ^{17}O and ^{18}O being arranged with known

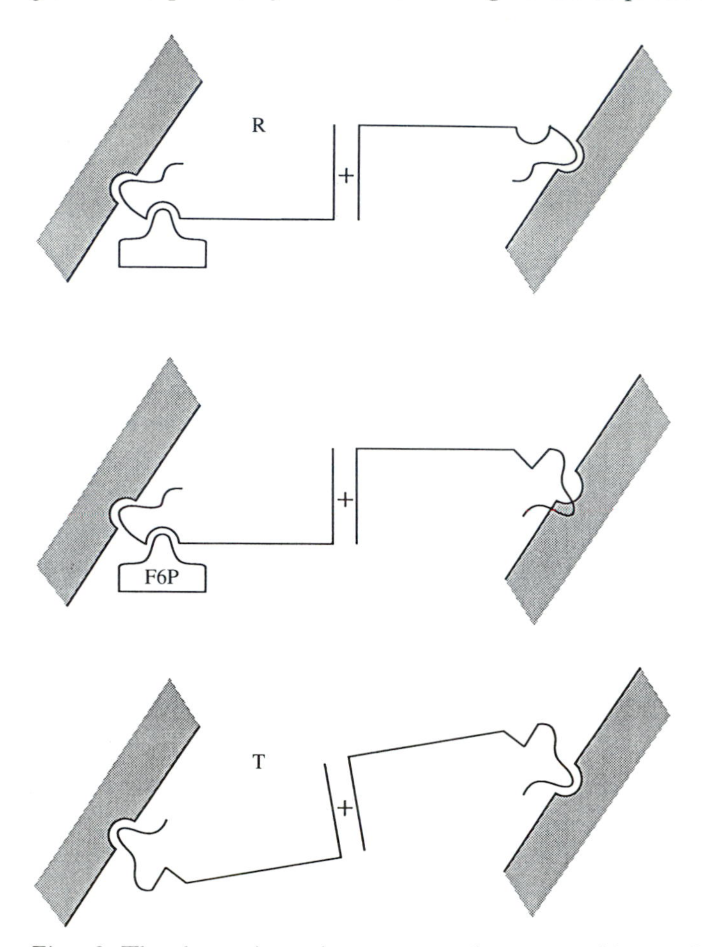

Fig. 38. The change in tertiary structure between residues 156 and 162, shown in Fig. 37, acts as a two-way switch, ensuring concerted transitions between alternative quaternary and tertiary structures. (From Schirmer & Evans, 1989.)

chirality in the γ-phosphate of ATP. If transfer from ATP to substrate is direct, then the chirality seen from the direction of the phosphorus atom is reversed; if it is via a covalently linked enzyme intermediate, it is preserved (Knowles, 1980). In PFK chirality was found to be reversed, whence transfer must be direct, probably by nucleophilic attack of the sugar 1-OH on the γ-phosphate leading to formation of a penta-coordinated phosphate transition state intermediate, followed by hydrolysis of the β-γ phosphate bond of ATP (Jarvest *et al.* 1981). The crystal structure shows that an asparate (127) lies close to the 1-OH of the sugar, which suggests that it polarizes the hydroxyl, drawing away its proton and making it into a strong nucleophile. Hellinga & Evans (1987) tested this hypothesis by directed mutagenesis of *E. coli* PFK. They replaced the aspartate by a serine and measured the kinetics of the mutant enzymes in the presence of saturating amounts of the activator GDP. This ensured that the kinetics of the forward reaction showed a hyperbolic rather than sigmoidal dependence on [F6P] or, in the reverse reaction, on [F1,6P_2]. The replacement of asparate 127 by serine reduced the catalytic rate in the forward reaction 18000-fold and in the backward reaction 3100-fold, suggesting that the aspartate acts as a general base, i.e. by attracting a proton in the forward reaction, and donating a proton in the backward reaction.

Lau & Fersht (1987) have used directed mutagenesis to find out how the effectors work. They selected three cationic side-chains that bind to the phosphates of the effectors and replaced them in turn by alanines. All three replacements made the binding of Mg-GDP and phosphoenolpyruvate either undetectable or much weaker – results that merely confirmed the binding site determined by X-ray analysis. On the other hand, replacement of glutamate-187 by alanine gave rise to unexpected and intriguing effects, hailed by the editors of *Nature* as being 'hard to reconcile with allosteric theory'. The replacement made the binding of Mg-GDP undetectable even at millimolar concentrations; it turned the inhibitor phosphoenolpyruvate into an activator, but the phosphoenolpyruvate-bound R-state had a V_{max} 40% lower than that of the wild type.

Since Lau & Fersht's paper was published, Schirmer & Evans (1989) found Glu 187 to be the one residue in the effector site that changes its conformation in the allosteric transition. We have seen that in the R-structure it binds to Mg^{2+} and to lysine 213; in the T-structure it turns away and takes the place of leucine 205. When effectors bind, residues 212–216 change their conformation and exert a leverage on the loop that extends to the active site. Replacement of glutamate 187 by alanine evidently disturbs that delicate mechanism in two ways. It alters the effector binding site so that the allosteric equilibrium of the phosphoenolpyruvate complex with the enzyme is shifted towards the R-state, and it alters the tertiary structure of the R-state, probably by shifting arginine 162 so as to reduce the catalytic activity.

Similar effects in haemoglobin mutants have often been described as hard to reconcile with Monod's allosteric model, but X-ray analysis has proved invariably that these haemoglobins alternate between the same two quaternary structures as the wild type, and that the apparent anomalies were due to changes in tertiary structure of the subunits that the normal two quaternary structures were able to accommodate. The same is likely to hold in phosphofructokinase and other allosteric proteins. For example, Eisenstein *et al.* (1989) engineered several different amino acid substitutions in the zinc-binding domain of aspartate transcarbamylase, the enzyme discussed in the next section. Most of them altered Hill's coefficient n, as was to be expected, but four out of five also altered V_{max}, and the changes in V_{max} bore no readily understandable relationship to the changes in n. One substitution reduced n to unity because it inhibited the $R \rightarrow T$ transition, yet it lowered V_{max}, while another substitution left n unchanged, but doubled V_{max}. Such apparently contradictory observations do not invalidate allosteric theory; on the contrary, when followed up by X-ray analysis they have always reinforced it.

In PFK as in haemoglobin, cooperativity and feedback inhibition arise from a transition between two alternative quaternary structures that rotates one pair of rigidly linked subunits relative to the other pair and rearranges the hydrogen bonds between them, but here the analogy ends. In no sense can the inactive PFK structure be described as tense and the active one as relaxed: the letters T and R are used merely for convenience. Another difference from haemoglobin is seen in the disposition of the active sites: in haemoglobin the effector sites lie at the subunits boundaries where the allosteric transitions take place, while the haems lie in pockets within the subunits, and the transmission of stereochemical effects between them and the effectors can only be inferred indirectly. In PFK, on the other hand, the catalytic sites span the subunit boundary where the allosteric transition takes place and are directly affected by it, while the effectors span a rigid subunit boundary and exert their influence through changes in tertiary structure that can be seen clearly and are comparable to the action of a set of levers (Fig. 38).

7. FEEDBACK INHIBITION OF A BIOSYNTHETIC PATHWAY: ASPARTATE TRANSCARBAMOYLASE

7.1 *Allosteric behaviour*

The study of feedback inhibition of biosynthetic enzymes began with Abelson's (1954) discovery that addition of isoleucine to the culture medium of *E. coli* inhibited the biosynthesis of isoleucine. Umbarger & Brown (1958) found this inhibition to be due to the specific action of isoleucine on the first enzyme in isoleucine biosynthesis, threonine deaminase. SH-reagents inactivated the inhibition without loss of enzymic activity, indicating that it was not competitive and that threonine and isoleucine bound to different sites (Changeux, 1961). That was a decisive observation, soon to be repeated in another biosynthetic enzyme, aspartate transcarbamoylase (ATCase) from *E. coli*, now officially renamed aspartate carbamoyltransferase. The enzyme is made up of two catalytic trimers (c_3) and three regulatory dimers (r_2). Each catalytic chain (c) has a molecular weight of 33000; each regulatory one (r) has a molecular weight of 17000 and contains one Zn^{2+} (Weber, 1968; Wiley & Lipscomb, 1968). The enzyme catalyses the first step in the biosynthesis of pyrimidines, the reaction of carbamoylphosphate with aspartate to form carbamoylaspartate, which is a precursor of the pyrimidine ring.

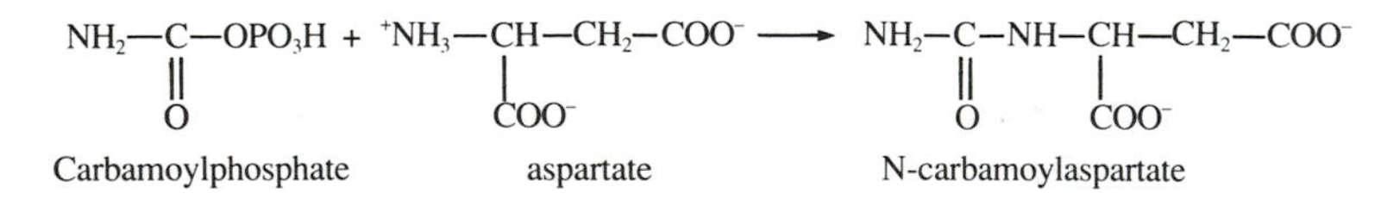

Using cell-free extracts of *E. coli*, Yates & Pardee (1956) found this reaction to be inhibited by cytidine and cytidine-5-phosphate. Crystallization of the enzyme by Shepherdson & Pardee (1960) allowed Gerhart & Pardee (1962) to study feedback inhibition for the first time in a pure enzyme. They discovered catalysis to be inhibited by CTP and activated by ATP, and they also demonstrated (without comment) the sigmoid dependence of the catalytic rate on the concentration of asparate. SH-reagents, urea, or heat, inactivated the inhibition but increased the catalytic activity. Gerhart & Pardee concluded that 'the surface

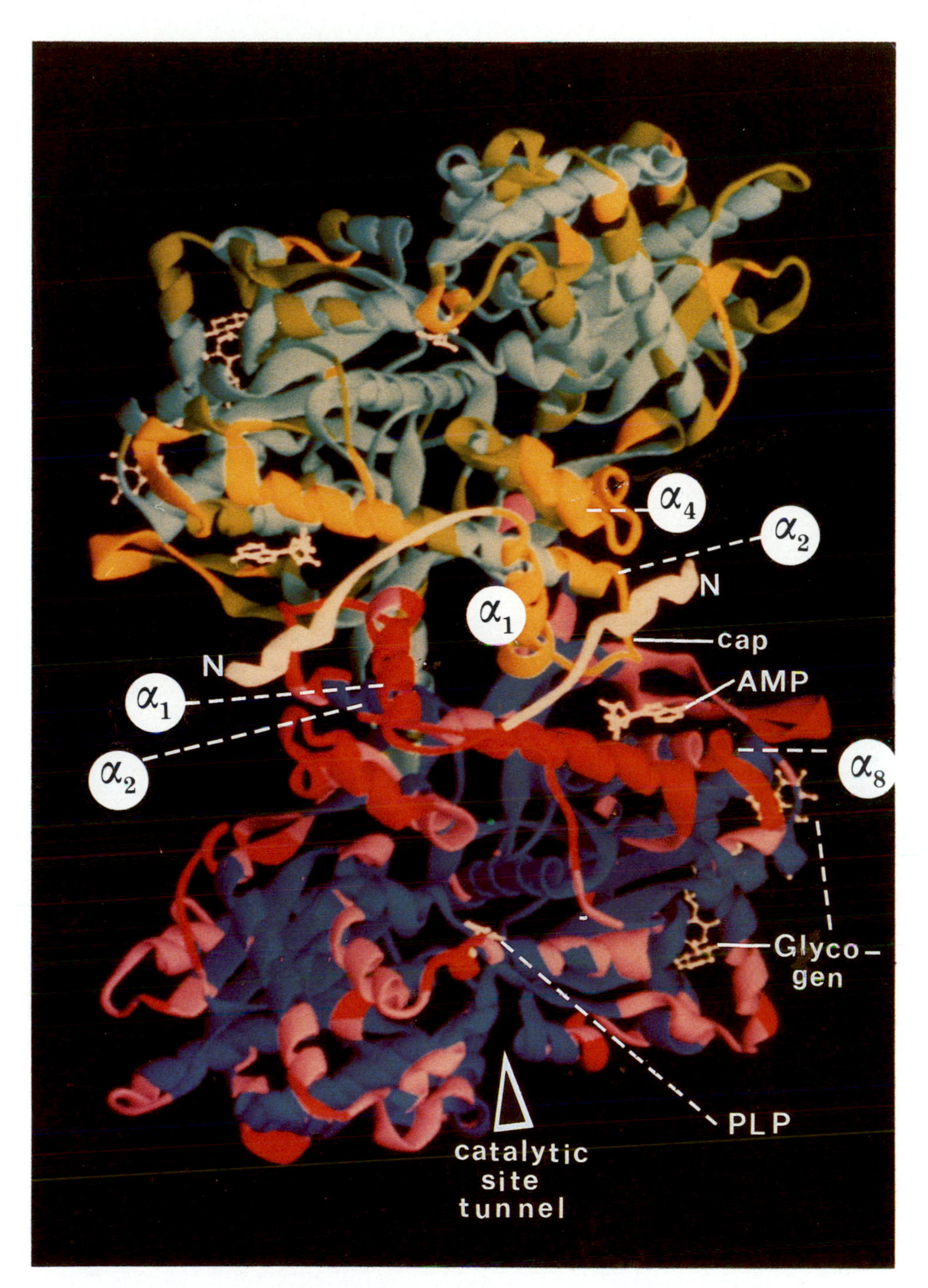

Fig. 25. Course of the polypeptide chains in a dimer of glycogen phosphorylase b in the T-structure, viewed along the axis of twofold symmetry, showing the tunnels leading to the catalytic pyridoxal phosphate (PLP) sites, the glycogen storage sites and AMP regulatory sites. Each subunit is colour-coded to show the difference in C_α-positions between phosphorylase b and a in the T-state. Shifts 0-0.4Å: blue (bottom subunit) and cyan (top subunit); 0.4-0.6Å: pink (bottom subunit) and green (top subunit); >0.6Å: red (bottom subunit) and yellow (top subunit). (courtesy Dr. R. J. Fletterick).

M. F. Perutz

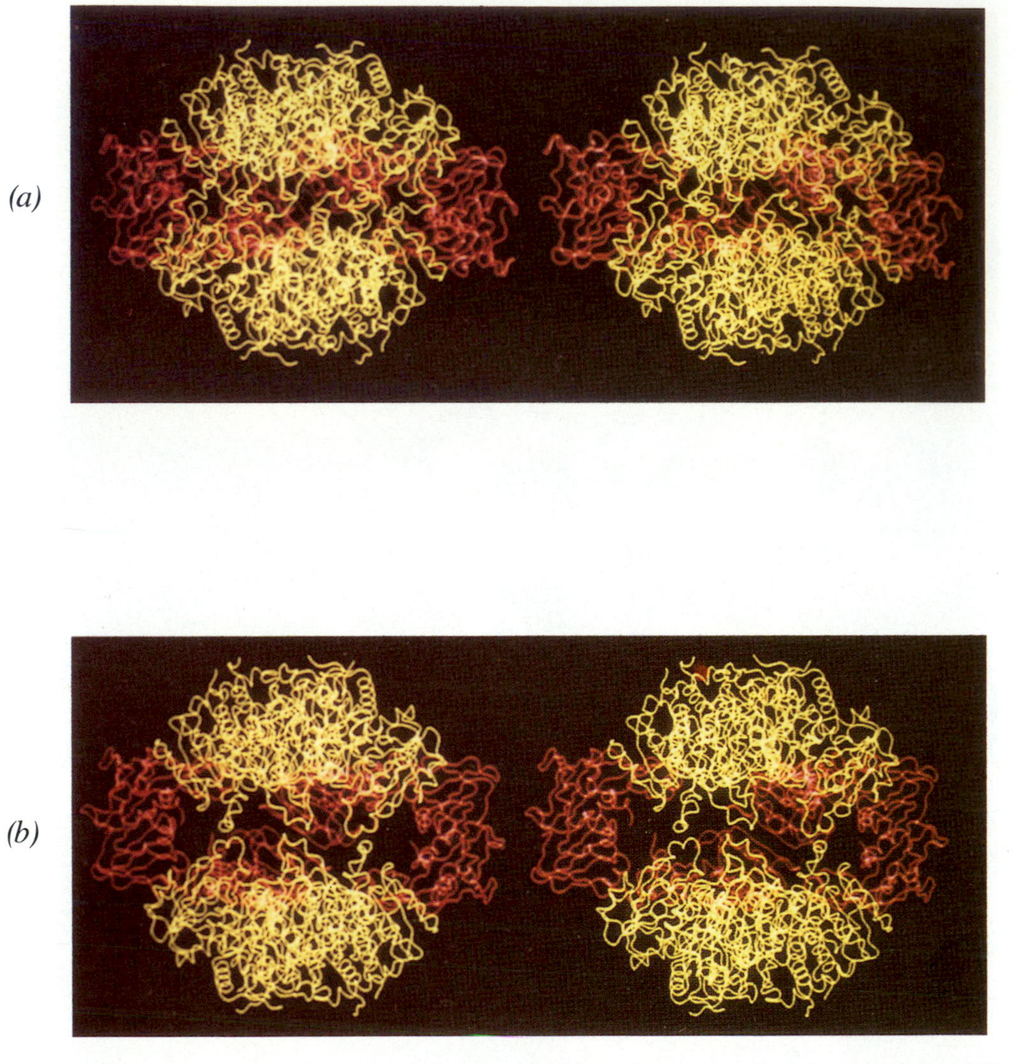

Fig. 41 a,b. α-carbon model of aspartate transcarbamylase viewed normal to the axis of threefold symmetry in (a) the T and (b) the R structure. The catalytic trimers are yellow and the regulatory dimers are red (courtesy Dr. W. N. Lipscomb).

(a)

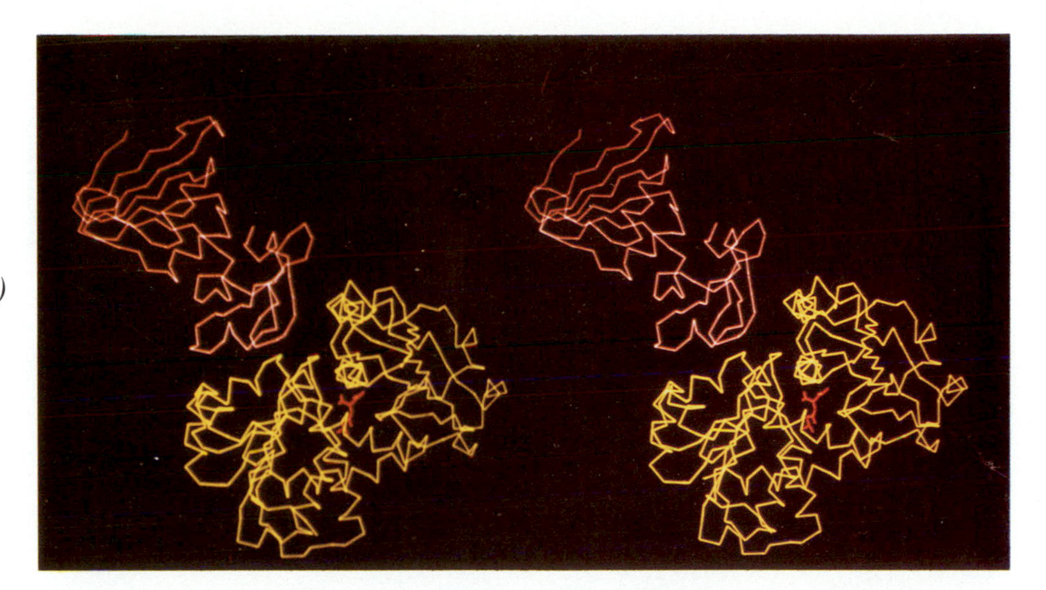

(b)

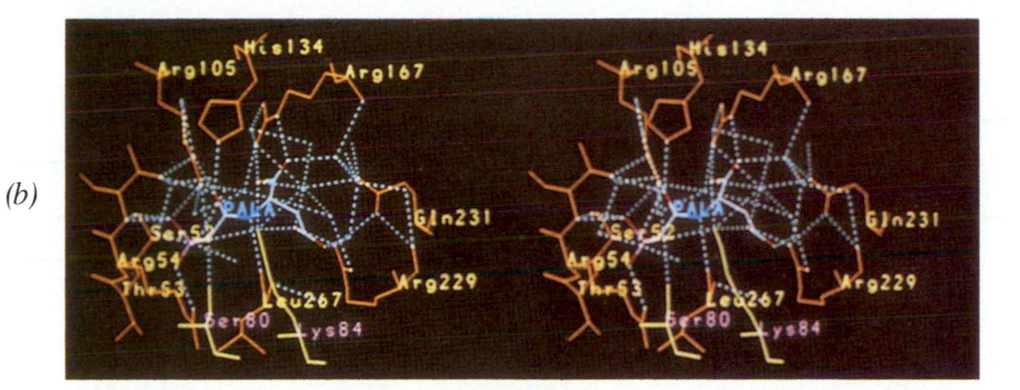

Fig. 44 a,b. Catalytic site of aspartate transcarbamylase. (a) Regulatory subunit, magenta, and catalytic subunit, yellow, showing the bisubstrate analogue PALA in red, occupying the cleft between the N- and C-terminal domains. The regulatory subunit has its Zn-binding C-terminal domain nearest to the catalytic subunit. The N-terminal domain with its extensive β-pleated sheet combines with effectors (courtesy Dr. Arthur Lesk). (b) Close up of PALA in the active site. Residues from the catalytic subunit in (a) are shown in yellow. Ser 80 and Lys 84 from the neighbouring subunit in the catalytic trimer are shown in magenta (courtesy Dr. W. N. Lipscomb)

(a)

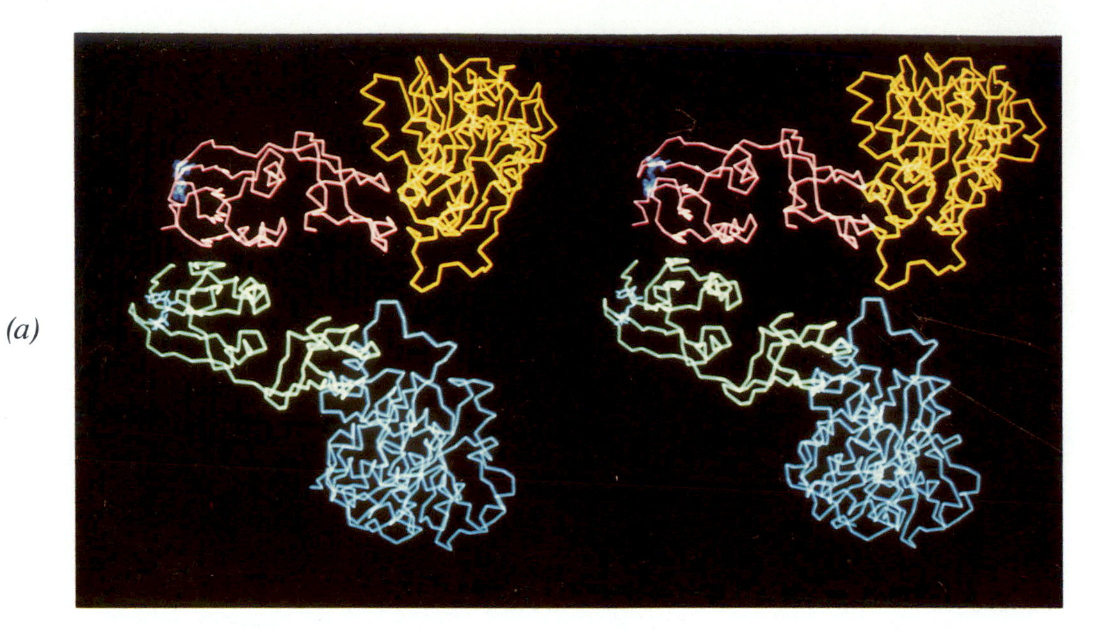

(b)

Fig. 46 a,b. Regulatory dimer (green and magenta) attached to catalytic subunits from different trimers (yellow and blue) in (a) the T-structure showing CTP (blue) in the regulatory sites at the tips of the N-terminal domains and (b) the R structure showing PALA (red) in the catalytic sites. Note that the contact between the regulatory subunits is the same in R and T. The views are approximately normal to the molecular triad and to one of the molecular dyads (courtesy Dr. Arthur Lesk).

(a)

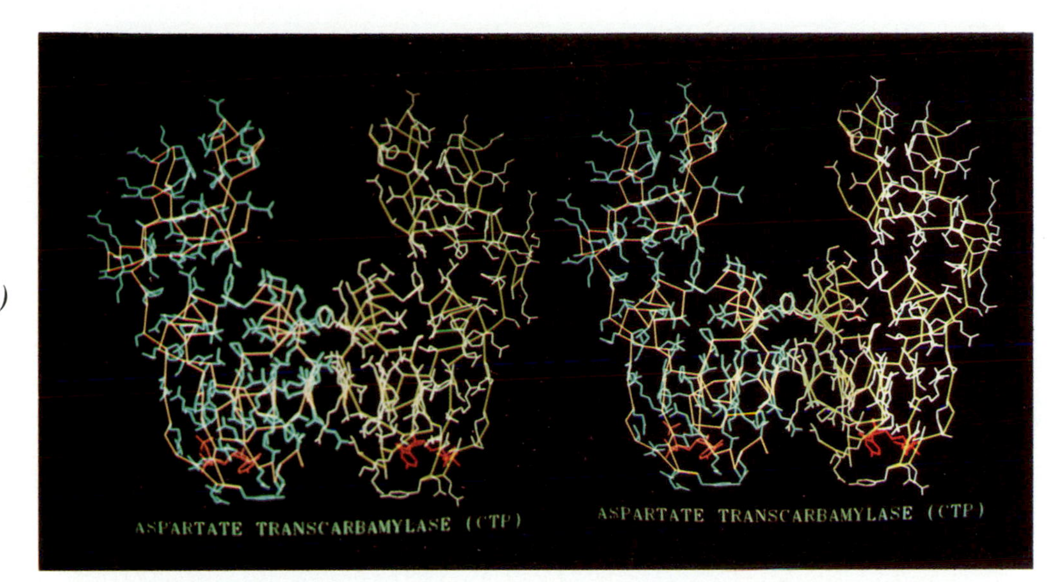

(b)

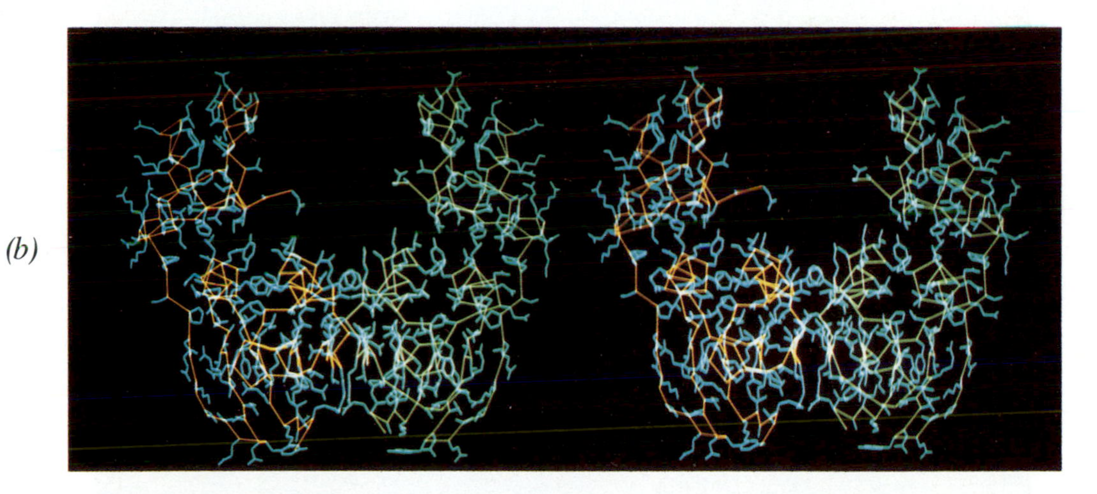

Fig. 47 a,b. Regulatory dimer turned by 90° from figure 46, enlarged and with side chains in (a) the T-structure with CTP in red, and (b) the R-structure. Note that the contact between the two subunits is unchanged and that the binding of CTP induces a rotation of the C-terminal domain relative to the N-terminal one that moves the two domains closer together (courtesy Dr. Arthur Lesk).

(a)

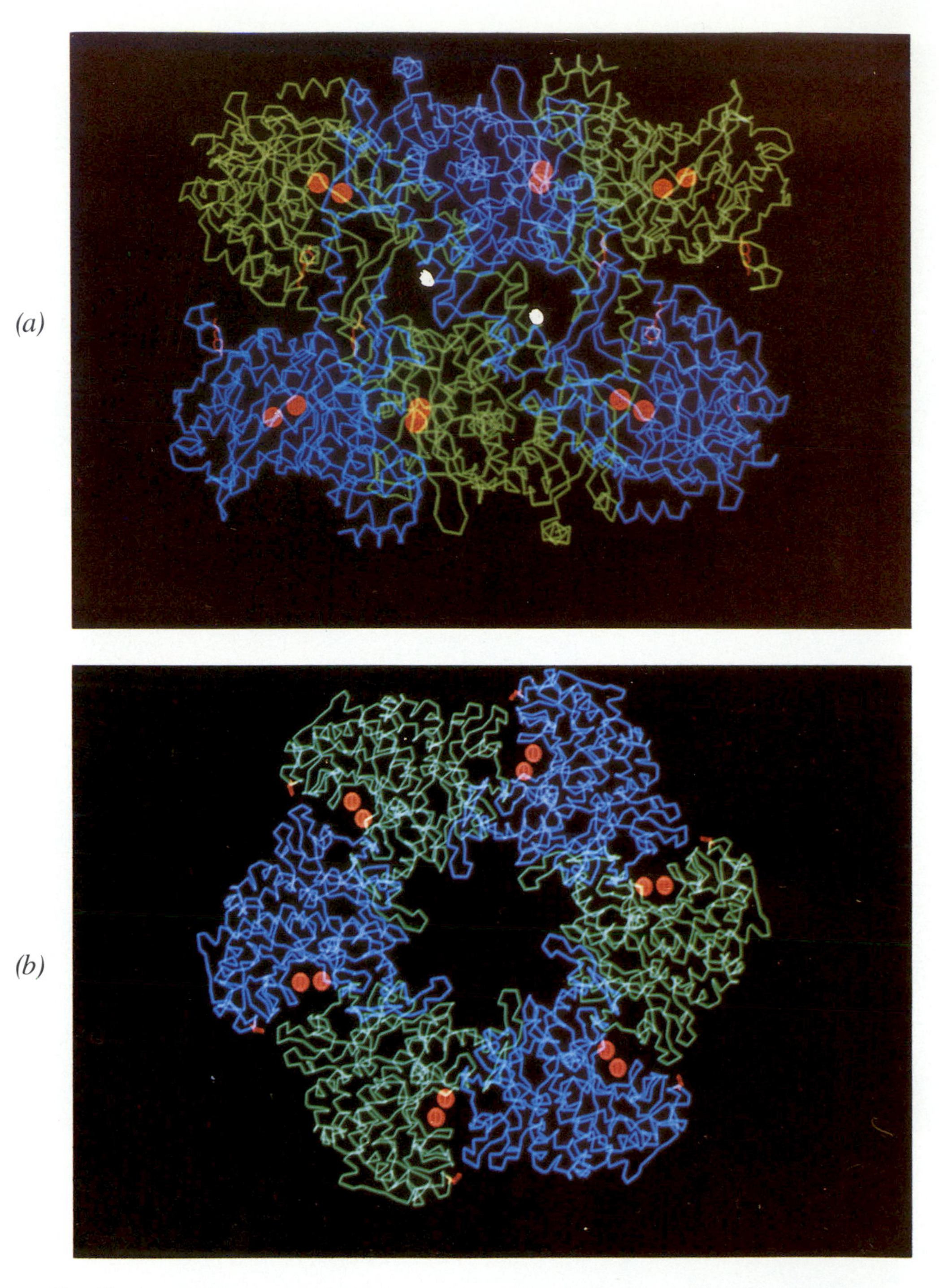

(b)

Fig. 49 a,b. α-carbon trace of glutamine synthetase from Salmonella typhimurium showing (a) two rings of six subunits normal and (b) one ring of six parallel to the hexad axis of symmetry. The red circles are Mn^{2+} ions in the active sites at the subunit boundaries. The white spots in (a) mark hydrophobic α-helices from one hexamer protruding into hydrophobic shafts in the neighbouring hexamer (courtesy Dr. David Eisenberg).

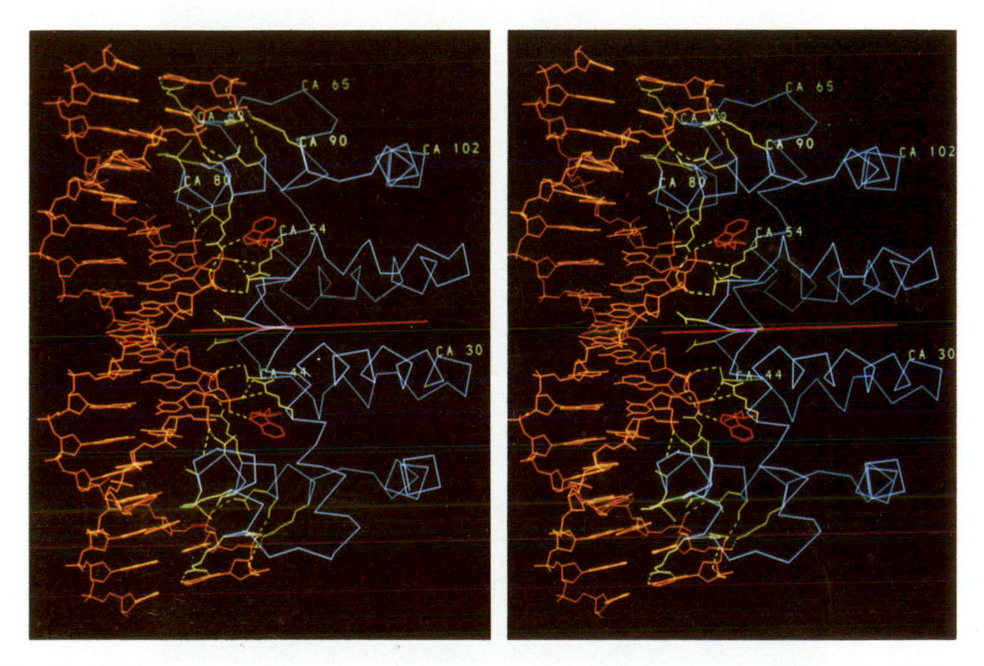

Fig. 52. The ternary complex of trp repressor, tryptophan (red) and operator DNA, showing the bending of the DNA in response to the hydrogen bonds linking donor groups of the repressor to phosphate oxygens. Note the marked propeller twist of the bases in the binding regions (from Otwinowski et al 1988). The operator used has the base sequence shown below.

```
T G T A C T A G T T A A C T A G T A C
  . . . . . . . . . . . . . . . . . .
  . . . . . . . . . . . . . . . . . .
  . . . . . . . . . . . . . . . . . .
  A C T G A T C A A T T G A T C A T G T
```

(a)

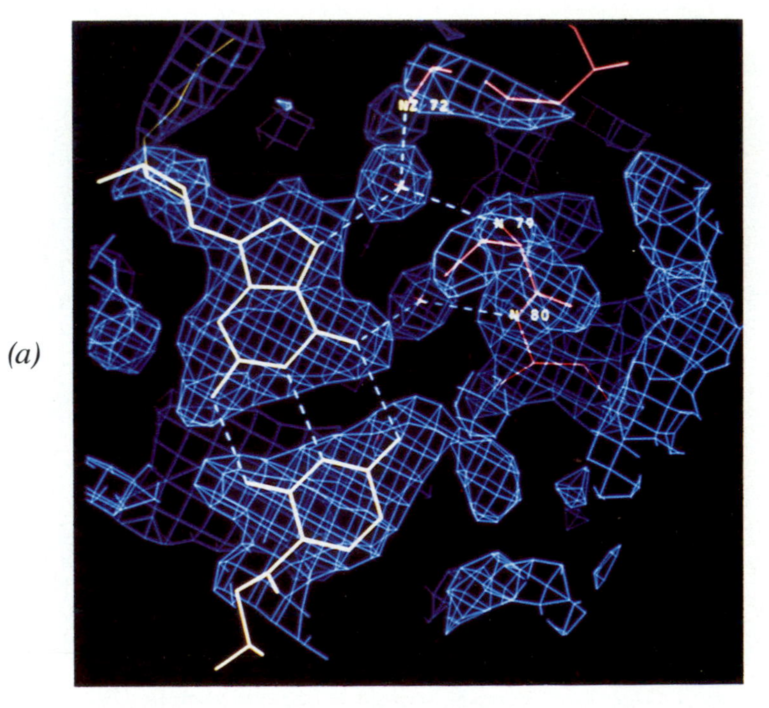

Fig. 53 (a). Electron density map showing hydrogen bonds between guanine 6, Lys 72 and other residues of the repressor mediated by two water molecules.

of the enzyme contains an exclusive site for binding CTP, distinct from the active site, but influencing the function of the latter'.

After formulating the theory of allosteric interactions together with Monod and Wyman, Changeux joined forces with Gerhart and Schachman to study aspartate transcarbamoylase as a prototype of an allosteric enzyme. Together they discovered the cooperative binding of a substrate analogue, succinate; the antagonistic effect of CTP on succinate binding; and the first experimental evidence that the conformation of ATCase really does change in response to activation or inhibition as predicted by the allosteric theory: addition of succinate and of the substrate carbamoylphosphate produced a sixfold rise in SH-reactivity and a drop of 3·6% in the sedimentation coefficient, indicative of an expansion or change of shape of the enzyme; these changes were opposed by CTP. They were complete at concentrations of substrate that only partially saturated the specific binding sites of the enzyme, which argued in favour of a two-state model (Changeux *et al.* 1968; Gerhardt & Schachman, 1968; Changeux & Rubin, 1968).

Whatever doubts may have remained about this conclusion have now been dispelled by a rigorous analysis of the sedimentation boundary of aspartate transcarbamoylase in equilibrium with the bisubstrate analogue *N*-(phosphonoacetyl)-L-aspartate (PALA), which binds to the enzyme very tightly (Werner & Schachman 1989; Werner *et al.* 1989).

$$O_3P{-}CH_2{-}\overset{\overset{\displaystyle O}{\|}}{C}{-}NH{-}\overset{\overset{\displaystyle COO^-}{|}}{C}H{-}CH_2{-}COO^-$$

The width of the boundary in the ultracentrifuge tube is determined by the differences in diffusion coefficients of the different species contained in the sample and by their rates of interconversion. Two distinct conformations of a protein interconverting slowly give a more widely spread boundary than the same two conformations interconverting fast, or a mixture of intermediate conformations. Since PALA binds to the enzyme very tightly, free and PALA-bound enzymes interconvert slowly. Werner & Schachman first calibrated the boundary spread produced by a mixture of separate non-interconverting enzymes in the R and T states: they spun equimolar quantities of a mutant enzyme that maintains the T-state even in the presence of an excess of PALA, and of wild-type enzyme that is completely converted to the R-state by an excess of PALA. They then titrated the sedimentation coefficient and boundary spread of wild-type enzyme alone with PALA. The sedimentation coefficient rose linearly until 4 mol of PALA per mol enzyme had been added and then remained constant at a value 3·5% higher than in the absence of PALA, as it had done on addition of an excess of succinate and carbamoylphosphate. The spread of the boundary, on the other hand, increased steeply and reached a maximum when no more than 2 mol of PALA per mol enzyme had been added, and then fell equally steeply to remain constant after 4 mol PALA per mol enzyme had been added. At its maximum, the spread of the boundary was the same as that of the equal mixture of enzyme in the R and T states employed for calibration, and broader than the boundaries calculated for mixtures of intermediate quaternary structures. At its minimum it corresponded to pure R-state, which has a boundary spread below that of the T-state, due to a reduction of the diffusion coefficient by the swelling in the T → R transition.

PALA had a striking effect on the reverse reaction catalysed by the enzyme, the arsenolysis of carbamoyl-L-aspartate. With this substrate bound, the enzyme remains predominantly in the T-state. The reaction requires high enzyme concentrations, because its rate is very slow (0·21 μM s^{-1}). PALA accelerates it. At [PALA/enzyme] < 0·05 that accleration is a linear function of [PALA], showing that it is caused by the binding of one PALA per enzyme molecule. The slope of the line corresponds to the formation of two catalytic sites in the R-state per PALA bound. This could mean either that two of each enzyme molecule's vacant sites are converted to R and the other three remain in T, or that 40% of the enzyme molecules complexed with PALA have the R and 60% the T-structure. The influence of allosteric effectors on the slope of the line strongly favoured the latter interpretation (Foote & Schachman, 1985).

In the absence of ligand, ATCase has 30 reactive SH groups; 6 in the catalytic and 24 in the regulatory subunits. Combination with PALA blocked the catalytic ones and raised the reactivity of the regulatory ones. When no more than half the catalytic sites were occupied by PALA, the reaction became biphasic, with one quarter of the SH groups reacting slowly and three-quarters reacting fast. All these and several similar observations were interpreted in terms of an equilibrium between a constrained or low affinity (T) state and a relaxed, or high-affinity (R) state of the enzyme, with values of $L = [T]/[R]$ of 250 in the absence of any ligands, 7 in the presence of carbamoylphosphate, 1250 in the presence of CTP without substrates, and 70 in the presence of ATP without substrates. The authors suggested that the enzyme undergoes concerted transitions between these two states which are triggered by the binding of different ligands to either the catalytic or regulatory subunits (Howlett & Schachman, 1977; Blackburn & Schachman, 1977; Howlett *et al.* 1977).

Past studies of the cooperative interactions of aspartate transcarbamoylase have been based on kinetics, but these are difficult to interpret without knowing the difference in the binding constants of the substrates to the R and T states of the enzyme. Newell *et al.* (1989) have measured the difference in the binding constant of PALA to the enzyme under three different conditions by equilibrium dialysis, using radioactively labelled PALA at nanomolar concentrations of enzyme and PALA. They obtained sigmoid binding curves with Hill's coefficients of 1·95 for the free enzyme, 1·35 with ATP and 2·27 with CTP. ATP reduced the average dissociation constant from the enzyme by the free energy equivalent of 1·9 kcal mol^{-1} and CTP raised it by the equivalent of 3·1 kcal mol^{-1}. The corresponding ratios of the L values quoted in the preceding paragraph give free energy changes of only 1·3 and 0·9 kcal mol^{-1}, showing that ATP and CTP affect not only L, but also the dissociation constants of PALA from the T and R structures, just as the heterotropic ligands of haemoglobin affect K_T as well as L.

Schachman's most striking evidence in favour of concerted transitions between alternative quaternary structures came from experiments with reconstituted

Table 2. Kinetic properties of hybrids and related species

Species	Active sites	Turnover number* $\times 10^{-3}$ (min^{-1})	K_M† (mM)	n_H‡	Inhibition by CTP§	Activation by ATP‖ (%)
$C_NC_NR_3$	6	9·9	7	1·75	80	158
$C_{nnp}C_{nnp}R_3$	4	8·0	11	1·37	—	—
$C_NC_PR_3$	3	9·4	12	1·38	75	122
$C_{npp}C_{npp}R_3$	2	8·5	15	1·34	75	152
$C_{nnp}C_PR_3$	2	7·5	15	1·36	75	152
$C_{npp}C_PR_3$	1	11·1	18	1·25	68	130
$C_PC_PR_3$	0	—	40	1·16	68	156
C_N	3	15·7	7	1·00		
C_{nnp}	2	13·6	7	1·00		
C_{npp}	1	16·3	8	1·00		
C_P	0		8	1·00		

* Units are number of carbamoyl aspartate molecules formed per minute per active site; corrections were made for the contribution of the pyridoxylated sites.

† Aspartate concentration corresponding to half maximal velocity.

‡ Hill coefficient corresponding to activity between 10 and 50% of V_{max}.

§ CTP concentration of 0·5 mM; aspartate concentration extrapolated to zero.

‖ ATP concentration of 2 mM; aspartate concentration extrapolated to zero.

hybrid molecules containing mixtures of native catalytic subunits (C_n) and subunits inactivated by reaction with pyridoxyl-5′-phosphate (C_p). These reconstituted molecules were all hexamers made up of two catalytic trimers and three regulatory dimers. Hybrids contained either one native and one pyridoxylated trimer, $(C_NC_P)R_3$; or trimers each containing two native and one pyridoxylated subunit, $(C_{nnp})_2R_3$; or one native and two pyridoxylated ones, $(C_{npp})_2R_2$; or two native subunits in only one of the trimers, $C_{nnp}C_PR_3$. All these hybrids exhibited cooperativity of the catalytic rate, inhibition by CTP and activation by ATP. V_{max} was a linear function of the number of active subunits in the hexamer, but the degree of inhibition or activation was independent of it (Table 2). Binding of carbamoylphosphate and succinate to the native subunits of the hybrids reduced the sedimentation constant and raised the rate of the reaction with paramercuribenzoate nearly as much as their binding to the fully native enzyme. These experiments proved that a change of tertiary structure of only two subunits, no matter whether they are in the same or in different trimers, can trigger a concerted change of the hexamer; similarly binding of CTP or ATP to the regulatory subunits can trigger such a change, no matter whether the regulatory subunits are linked to an active or inactive catalytic subunit. Native catalytic trimers alone, or compounds of less than six native catalytic and regulatory subunits, were non-cooperative (Gibbons *et al.* 1974; Gibbons *et al.* 1976), but a species c_6r_4, lacking one regulatory dimer, exhibited cooperativity of the catalytic rate and sensitivity of that rate to GTP and ATP (Evans *et al.* 1975). For a recent review of these and other allosteric properties of aspartate transcarbamylase see Schachmann (1988); Hervé (1989).

The experiments described so far attributed the regulatory effects of CTP and ATP chiefly to their influence on the allosteric equilibrium. Hervé *et al.* (1985) have challenged this view by examining the low-angle X-ray scattering of concentrated solutions of aspartate transcarbamoylase. Having found distinct scattering curves for the enzyme in the R- and T-states, they measured the influence of CTP and ATP on the scattering from solutions partially saturated with PALA (0·6 M PALA/M active site). Without effectors such solutions gave a scattering curve intermediate between R and T. CTP shifted the curve slightly towards T, while ATP had no effect. Hervé *et al.* concluded that ATP activates the enzyme, not by its influence on the allosteric equilibrium, but purely by direct transmission of a stereochemical change from the allosteric to the 50 Å-distant catalytic site.

I do not believe that this conclusion is correct, because the authors have failed to take account of the cooperative nature of the enzyme's reaction with PALA. This would have ensured that their solutions contained predominantly molecules that are either free from PALA or fully saturated with it. It is known that ATP does not switch the enzyme from T to R in the absence of substrate; hence it would not have affected the allosteric equilibrium of the fraction of enzyme that was free from PALA; the fully PALA-bound fraction was already in the R-state. Similarly CTP would not have affected the allosteric equilibrium of the fully PALA-bound fraction because PALA would have maintained the enzyme in the R-state regardless of CTP, nor would it have affected the PALA-free fraction that was already in the T-state. What about the partially PALA-bound fraction? According to Schachman's findings, this has predominantly the R-structure and would therefore have been unaffected by ATP; it would have been switched to the T-structure by CTP. The small CTP-induced shift of the curve would have been a measure of that fraction's concentration. In summary, the results can be readily explained without invoking the sterochemically improbable transmission of *purely* tertiary effects from the allosteric to the catalytic site, though such effects do exist side by side with quaternary ones.

7.2 *The three-dimensional structures and the mechanism of cooperativity*

The stereochemical mechanisms underlying these remarkable properties emerged after Lipscomb and his colleagues had solved the structure of the R-form in the presence of PALA or of carbamoylphosphate plus succinate; and that of the T-form in the presence of CTP and also in the absence of ligands (Honzatko *et al.* 1982; Honzatko & Lipscomb, 1984; Krause *et al.* 1987; Ke *et al.* 1988; Gouaux & Lipscomb, 1988; Kantrowitz & Lipscomb, 1988).

The catalytic trimers form two equilateral triangles. In the complete enzyme one of the triangles is inverted and stacked on top of the other in a nearly eclipsed position. Each of the three regulatory dimers forms an external bridge between neighbouring catalytic subunits in adjacent trimers. The subunits are related by

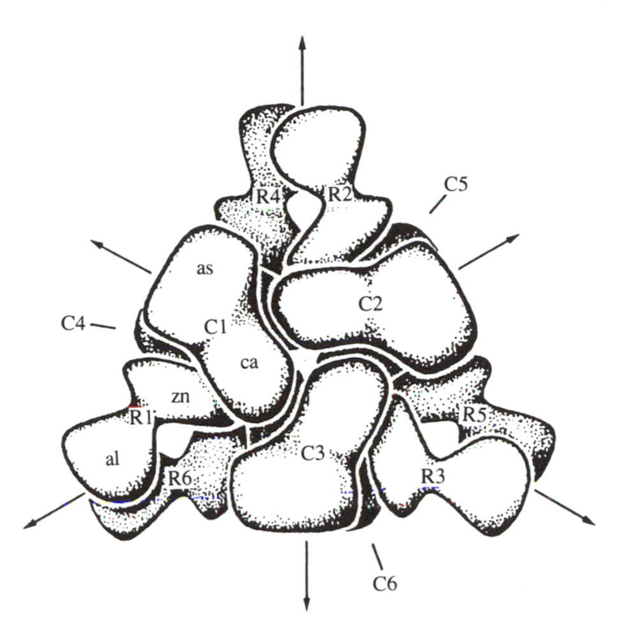

Fig. 39. Quaternary structure of aspartate transcarbamoylase of *E. coli*. Point group D_3 or 32. Catalytic trimers C_1—C_3 and C_4—C_6, with the N-terminal carbamoyl phosphate-binding domains (ca) and the C-terminal aspartate-binding domains (as); regulatory dimers with the N-terminal effector-binding domain (al) and the C-terminal Zn-binding domain (zn). (From Kantrowitz & Lipscomb, 1988.)

an axis of threefold symmetry that passes through the centre of the molecule, and by three axes of twofold symmetry perpendicular to the triad. (Fig. 39). The catalytic and regulatory subunits are each divided into two domains. The $T \rightarrow R$ transition induces a change of quaternary structure that resembles the action of a differential gear (Fig. 40). Suppose the horizontal gearwheels represented the catalytic trimers, and the vertical ones the regulatory dimers. Suppose further that the vertical gearwheels were elliptical. Then a concerted rotation of the vertical gearwheels would turn the two horizontal gearwheels in opposite directions and change the vertical spacing between them. On transition from T to R, the regulatory dimers rotate clockwise, as seen from the surface, by 15°; this rotation causes one catalytic trimer to turn counterclockwise by 10° relative to the other trimer about the threefold symmetry axis, and to move 12 Å further away from its partner along that axis. This is the expansion responsible for the rise in sedimentation coefficient first seen by Schachman and his colleagues. In addition, the domains of the catalytic and regulatory subunits move relative to each other. The end-points of the transition preserve the symmetry of the molecule. The contacts between the subunits making up the catalytic trimers and those between the subunits of the regulatory dimers undergo little change, while the contacts between the catalytic trimers and those between the regulatory and catalytic subunits change drastically (Fig. 41).

Each catalytic subunit is made up of two domains that enclose the active site in a pocket between them. The stereochemical changes that activate the catalytic site

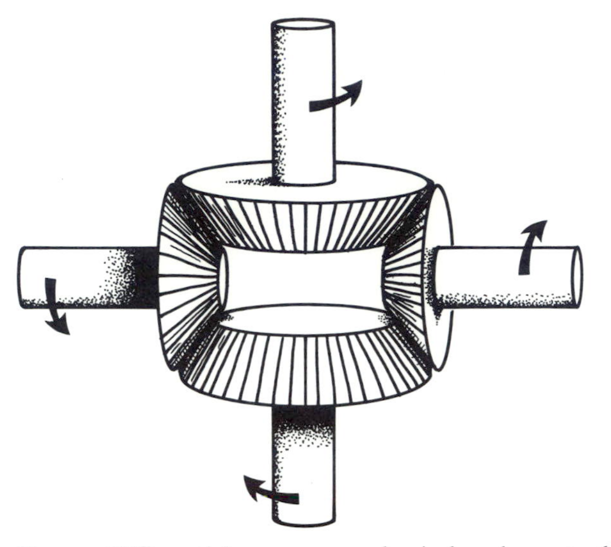

Fig. 40. Differential gear as a mechanical analogue to the enzyme aspartate transcarbamoylase.

on transition from T to R are compounded from rearrangements of segments belonging to three distinct subunits. Two long loops, the '80' and the '240' loops, protrude from each catalytic subunit. The '80' loops link the catalytic trimer around the threefold symmetry axis; each loop forms bonds with both its neighbours in both the T and R structures; in the T-structure these bonds keep serine 80 and lysine 84 away from the active site (Fig. 42), but on transition to the R-structure the serine and lysine become part of the catalytic site of the neighbouring subunit *within the same trimer*. In the T-structure the '240' loops tie *adjacent trimers* to each other: glutamate 239 forms hydrogen bonds with lysine 164 and tyrosine 165 of the adjacent trimer. The loop is also stabilized by an internal hydrogen bond between tyrosine 240 and aspartate 271. In the R-structure the trimers are pulled away from each other so that their contacts become more tenuous; many bonds break and the '240' loop folds back so that most of it becomes part of the catalytic site of its own subunit (Fig. 43). The construction of the contacts between the subunits ensures that the quaternary transition is concerted: it is not possible to change one of the contacts without changing the two others. The contraction of the '240' loops and the widening gap between the catalytic trimers explain the rise in the sedimentation constant of catalytic trimers on addition of PALA.

The catalytic site is in a cleft lined by segments of chain from the N-terminal and C-terminal domains and the '80' loop from the neighbouring subunit in the same trimer (Fig. 44). In the reaction with the natural substrates the N-terminal domain forms the carbamoyl phosphate binding site and the C-terminal domain forms the aspartate binding site. In the T-structure the aspartate and carbamoylphosphate binding sites are too far apart for the substrates to react and the '80' loop is out of their reach. In the R-structure the gap between the two

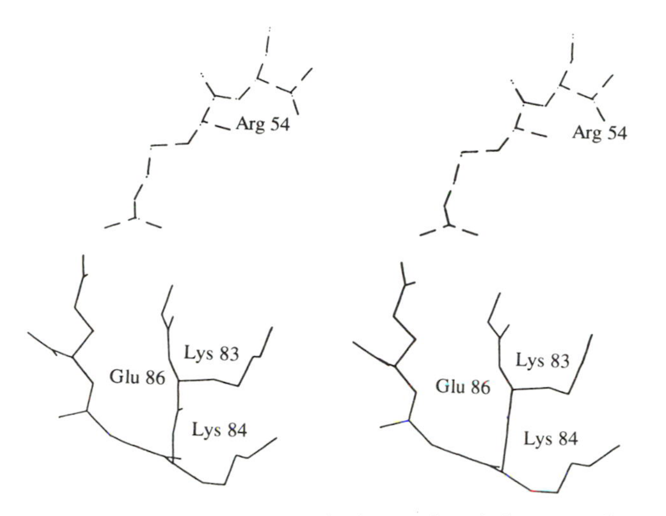

Fig. 42. In the T-structure hydrogen bonds between Arg 54 of one catalytic subunit and Glu 86 of a neigbouring catalytic subunit *within* the trimer tie down Lys 84 and prevent its participation in substrate binding, as shown in Fig. 44. (From Honzatko *et al.* 1982.)

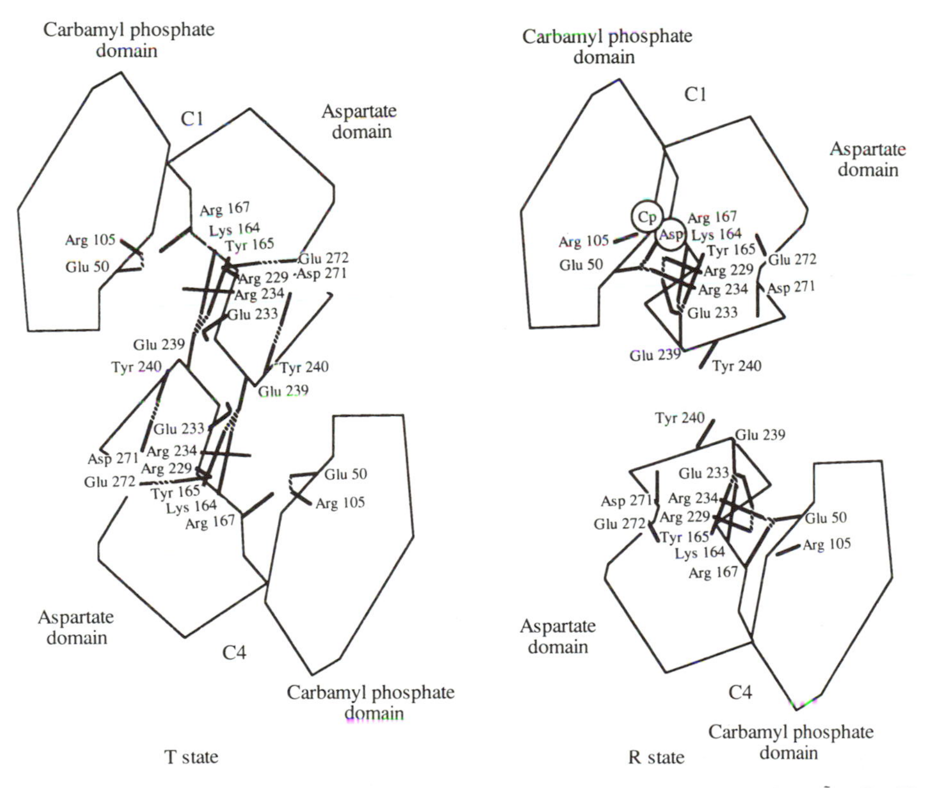

Fig. 43. Two catalytic subunits of neighbouring trimers in the T and R states. In T, the '240 loops' are held by hydrogen bonds linking neighbouring trimers. In the R-state, these bonds are broken, the loops fold back to become part of the aspartate binding sites and the gap between the two domains of each catalytic subunit narrows. The trimers move apart. (From Kantrowitz & Lipscomb, 1988.)

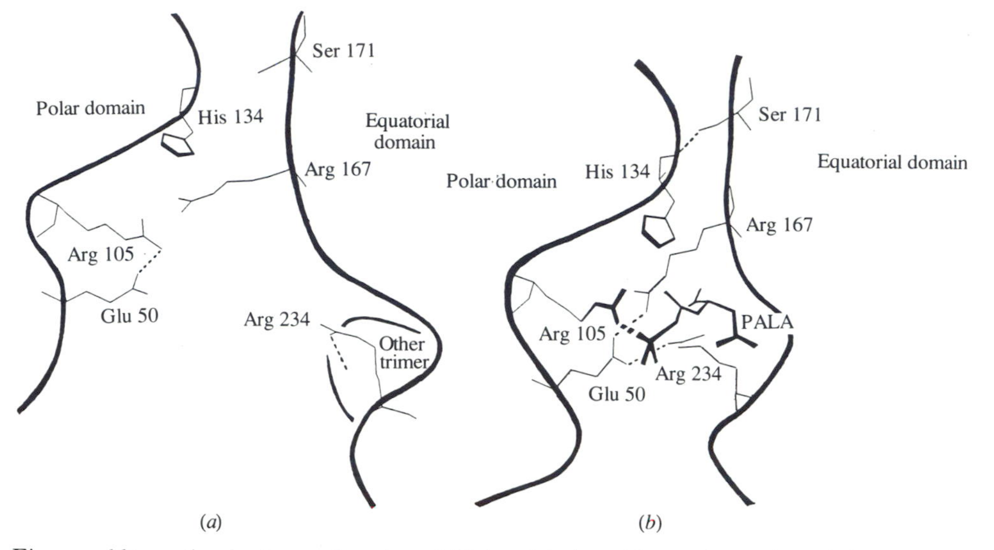

Fig. 45. Alternative hydrogen bonds stabilizing (*a*) the wide conformation of the active site in the T-structure, and (*b*) the narrow one in the R-structure of aspartate transcarbamoylase. (Courtesy Dr W. N. Lipscomb.)

domains narrows by 3 Å, and hydrogen bonds link glutamate 50 from the N-terminal domain to arginines 167 and 234 of the C-terminal one, thus fashioning and reinforcing the catalytic site, together with the '80' loop from the neighbouring subunit (Fig. 45). The transition state analogue PALA forms hydrogen bonds with residues of all these segments as shown in Fig. 44(*b*).

The regulatory subunit also consists of two domains, an N-terminal one that carries the effector binding site and the contact with the other regulatory subunit in the dimer, and a C-terminal one that makes contact with two catalytic subunits in the T-structure and with only one in the R-structure. The C-terminal domain contains an atom of zinc that plays a purely structural role. In the allosteric transition each regulatory subunit undergoes a marked change in tertiary structure. The significance of this change is seen in Fig. 46, which shows the movement of the C-terminal domains of two regulatory subunits and of the catalytic subunits attached to them relative to their N-terminal domains. It illustrates how this movement acts as a lever, keeping the catalytic subunits apart in the R-structure and pulling them together in the T-structure. Fig. 47 shows another view of the two regulatory chains in the T and R structures. The contact between the two effector-binding domains hardly changes; the main change consists in a reorientation of the two C-terminal domains. Note also that CTP binds to an external pocket in the N-terminal domain. It is still unclear how this sets the large tertiary changes in motion.

The location of the active site and the participation in catalysis by residues from neighbouring subunits in the trimer have been confirmed by directed mutagenesis. Serine 52 of one catalytic subunit binds to the phosphate group of PALA, and lysine 84 from the neighbouring catalytic subunit binds to both the phosphate and to one of the carboxylates of PALA (Fig. 44*b*). Replacement of

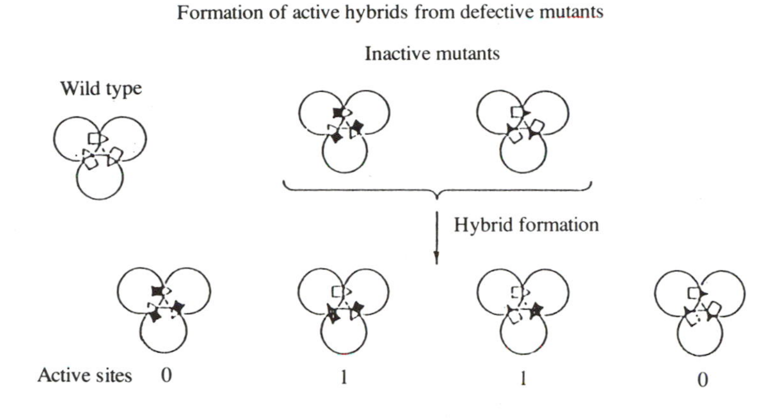

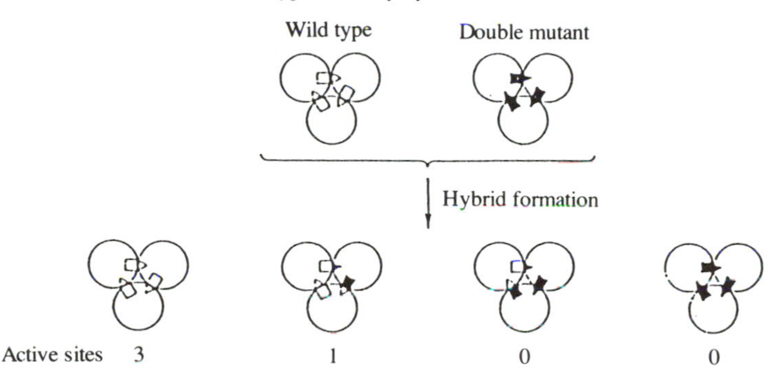

Fig. 48. Formation of active hybrid aspartate transcarbamoylase trimers from catalytically inactive subunits having either one of two different substitutions in the active site, marked by black squares and black triangles. Empty squares and triangles mark wild-type residues. (From Wente & Schachmann, 1987.)

either the serine by histidine or the lysine by glutamine therefore inactivated the enzyme. On the other hand, hybrid molecules made up of catalytic trimers containing either two subunits with His 52 and one subunit with Glu 84, or one subunit with His 52 and two subunits with Glu 84, showed one-third of the catalytic activity of the wild type. This proved that active catalytic sites can be reconstituted from one subunit containing Ser 52 and another containing Lys 84 (Fig. 48). In this way interallelic complementation could produce a viable organism from two alleles, neither of which is viable on its own (Fincham & Day, 1963; Wente & Schachmann, 1987).

Crick & Orgel (1964) proposed a theory of interallelic complementation. They suggested that in a dimer, composed of two inactive monomers A and B, activity could be restored if correct folding of the chain in position a of monomer A rectified the misfolding of the chain in position b of monomer B, and correct folding of the chain in position c of monomer B corrected misfolding of the chain in position d of monomer A. For once, the real mechanism turns out to be simpler than the hypothetical one.

In the abnormal human haemoglobin Kansas, disruption of a pair of hydrogen bonds that stabilize the R-structure reduced the oxygen affinity and raised the allosteric constant L (see p. 19). Analogous replacements were made in aspartate transcarbamylase by directed mutagenesis. Replacement of arginine 234 by serine, or of glutamate 50 by glutamine broke a strong hydrogen bond that forms part of the scaffolding of the active site in the R-structure (Fig. 45*b*). In isolated catalytic trimers the first of these replacements raised K_M^{Asp} 45-fold and lowered k_{cat} 33-fold; the second one raised K_M^{Asp} and lowered k_{cat} only about 10-fold. The difference is accounted for by the electrostatic factor: the second replacement leaves in the active site a net positive charge that attracts aspartate, while the first one leaves a net negative charge that repels it. In the mutant holoenzymes aspartate concentrations above 0·1 M were needed to observe significant activity, and cooperativity was almost absent, suggesting a T-state. The addition of 0·1 mM PALA nearly doubled k_{cat} of the glutamine 50 mutant because PALA works as a powerful allosteric activator even though it blocks most of the catalytic sites. The mutant could also be weakly activated by replacing tyrosine 240 by phenylalanine; this additional replacement lowered L by disrupting part of the scaffolding of the T-structure (Fig. 43*b*). It even restored some of the lost cooperativity (Ladjimi *et al.* 1988; Ladjimi & Kantrowitz, 1988; Middleton & Kantrowitz, 1988). Here structural analysis and directed mutagenesis have complemented each other in a most elegant manner.

Glutamate 239 stabilizes the T-structure by its hydrogen bonds with lysine 164 and tyrosine 165 (Fig. 43, p. 61). Ladjimi & Kantrowitz (1988) replaced this key residue by glutamine and found the mutant enzyme to be non-cooperative with respect to aspartate; it also failed to be activated by PALA, indicative of being 'frozen' in the R-state. On the other hand, the substrate- and effector-free enzyme crystallized in an intermediate structure close to T: the catalytic trimers are 2 Å further apart. Single crystals could be converted to the R-form by immersion in 80 mM aspartate + 40 mM phosphate (W. N. Lipscomb, private communication). Low angle X-ray scattering from the mutant gave a larger radius than for the unligated wild-type confirming an expanded t-state. The mutant and wild-type radii were the same in the presence of PALA suggesting an unaltered R-structure (Tauc *et al.* 1989). Why should an intermediate structure close to T have formed in crystals and in concentrated solutions, but not have manifested itself in the dilute solutions used for the functional studies? The only explanation that comes to mind is dissociation, yet this seems unlikely in view of the many contacts between the catalytic and regulatory subunits.

8. CONTROL OF NITROGEN METABOLISM: GLUTAMINE SYNTHETASE

Glutamine synthetase catalyses the amination of glutamate by ammonia with energy provided by ATP. This is a key function, because the amide group of glutamine serves as a source of nitrogen for the synthesis of many metabolites, among them the carbamoyl phosphate which forms one of the two substrates of aspartate transcarbamoylase discussed in the preceding section; also tryptophan, histidine, glucosamine-6-phosphate, AMP and others. The bacterial enzyme is subject to feedback inhibition by all these metabolites as well as by alanine and glycine. Its susceptibility to inhibition is raised by adenylation of one of its tyrosines (397) by an adenylyl transferase which is itself subject to regulation by yet other metabolites. Inhibition of the enzyme by the different metabolites is cumulative rather than competitive. The enzyme requires divalent metal ions for activity. The adenylated enzyme is active only in the presence of Mn^{2+}, whereas the unadenylated enzyme is activated by Mg^{2+}, Ca^{2+} or Co^{2+}, but not by Mn (Stadtman & Ginsburg, 1974).

Eisenberg and his colleagues have solved the structure of glutamine synthetase from *Salmonella typhimurium* combined with manganese ions in its unadenylated form at 3·5 Å resolution (Almassy *et al.* 1986; Eisenberg *et al.* 1987; Yamashita *et al.* 1989). It is a dodecamer of identical subunits of molecular weight 51 628 which are stacked together in two rings of six placed face to face, so that subunits in adjacent rings are related by axes of twofold symmetry. Each subunit touches four of its neighbours (Fig. 49). The electron density map outlines the course of the polypeptide chain. Small side-chains are generally indistinguishable from each other at 3·5 Å resolution, but large ones like tryptophan and tyrosine can be identified, and the manganese ions appear as dominant peaks in the active sites which lie at the boundaries between adjacent subunits within the rings of six, in channels that run normal to the rings. The manganese ions are 20 Å away from tyrosine 397 whose adenylation sensitizes the enzyme to its inhibitors, but part of the lining of the active consists of a loop of chain with tryptophan 57 at its tip; this tryptophan is only 6 Å from tyrosine 397.

So far, this is the only one of the allosteric forms of the enzyme solved by X-ray analysis. Adenylation of tyrosine 397 did not affect the crystal lattice (Heidner *et al.* 1978). Several feedback inhibitors bound to different locations in the active site channel without producing significant structural changes. Addition of the partial transition-state analogue methionine-sulphoxime together with ATP to a solution of glutamine synthetase produced a methionine-sulphoxime-ADP complex. Crystals of this complex showed large changes in the diffraction pattern, indicative of structural changes, but these have not yet been elucidated (D. Eisenberg, personal communication). Fig. 4 suggests that the simplest allosteric transition would be a rotation of the two rings relative to each other about the sixfold symmetry axis, and/or a translation along that axis. The structure shows hydrophobic α-helices from one ring of subunits fitted into hydrophobic sockets of the opposite layer; two of these helices are marked by white dots in Fig. 49. If they were able to slide in their sockets like pistons in cylinders, then allosteric transitions could expand or contract the spacing between the two rings as in aspartate carbamoylase. If the enzyme alternated between only two quaternary structures R and T, then different inhibitors could all stabilize the same T-structure by combining with a variety of different sites, just as 2,3-diphosphoglycerate, bezafibrate and related compounds did by combining with different sites in the T-structure of haemoglobin. This would account for their cumulative rather than competitive action. On the other hand, there is no firm evidence yet for a change in quaternary structure.

In the presence of Co^{2+}, heavily adenylated glutamine synthetase polymerizes first to single-stranded filaments and them to helical cables (Frey *et al.* 1975). The helical pitch of these cables varies over wide limits owing to differences in the stacking of the individual molecules. These angular variations may reflect allosteric states with different angles of rotation of the two hexameric rings relative to each other, but they could also have other causes.

9. COOPERATIVITY AND FEEDBACK INHIBITION WITHOUT CHANGE OF QUATERNARY STRUCTURE: THE *trp* AND *met* REPRESSORS OF *E. coli*

Genetic and biochemical analysis of the repression of tryptophan biosynthesis in *E. coli* formed part of the research that led Monod & Jacob (1961) to formulate their operon hypothesis. Monod & Cohen-Bazire (1953) had found that 5-methyl tryptophan represses the bio-synthesis of tryptophan, even though it does not substitute for tryptophan in proteins, and that it inhibits the growth of wild-type *E. coli*. Cohen & Jacob (1959) then discovered that 5-methyl-tryptophan-resistant mutants arose by mutation of a gene (R_T) located far from the cluster of genes coding for the tryptophan-synthesizing enzymes, and they concluded that the R_T gene controls the rate of synthesis of several proteins through a cytoplasmic intermediate. The isolation of that intermediate posed formidable difficulties because of its low concentration: we now know that even in the presence of saturating amounts of tryptophan, one *E. coli* bacterium contains only 375 molecules of it, equivalent to a concentration of .85 μM (Gunsalus *et al.* 1986). It took nine years until Zubay and his colleagues (1972) enriched an extract of *E. coli* proteins in the repressor; shortly afterwards Rose *et al.* (1973) and McGeoch *et al.* (1973) demonstrated that addition of partially purified tryptophan repressor together with tryptophan blocks the initiation of transcription of the *trp* operon DNA in a cell-free system. Another ten years passed before the repressor was finally purified and crystallized from an overproducing strain of *E. coli* by Joachimiak *et al.* (1983). They found that the repressor binds tryptophan non-cooperatively with a dissociation constant of 20 μM. The repressor binds the operator with a dissociation constant of 0·2–0·5 nM; its dissociation constant from non-specific DNA was about a thousand times larger.

Carey (1988) determined the dissociation constant K_D of the repressor from operator DNA by retardation of the complex in a non-denaturing gel. She split a 490 base-pair fragment from the operator region of *E. coli* into a 415 base-pair fragment that contained the operator sequence and a 75 base-pair fragment that served as a control. The larger fragment formed repressor complexes with a K_D of about 0·5 nM, while the smaller one remained free until the repressor concentration approached 100 nM. At this concentration both fragments formed multiple repressor complexes. DNase footprinting at the low repressor concentrations showed that only the operator region of the 415 base-pair fragment, extending from -25 to $+4$ base pairs from the start of transcription, was protected by the repressor. The stoichiometry of the complexes in the gels was determined using ^{32}P-labelled DNA and ^{3}H-labelled repressor and found to be $1{\cdot}0 \pm 0{\cdot}18$ repressor dimer per DNA molecule. These experiments have established both the stoichiometry and the specificity of the reaction of the repressor with the operator.

Sigler and his colleagues have determined the structures of the repressor with and without tryptophan, of the repressor with tryptophan bound to the operator and of the repressor with a tryptophan analogue that acts as an inducer (Schevitz *et al.* 1985; Zhang *et al.* 1987; Marmorstein *et al.* 1987; Lawson & Sigler, 1988; Lawson *et al.* 1988; Otwinowski *et al.* 1988; for review see Marmorstein & Sigler, 1988).

The apo-repressor is a dimer of two identical polypeptides each containing 107 amino acid residues. Each half-molecule forms the helix-turn-helix motif previously found to be the reading-head in other DNA-binding proteins (Ptashne, 1986). In addition, each subunit contains an α-helical core that intertwines with the symmetry-related core of its partner subunit. This core is rigid and largely unaffected by the binding of either tryptophan or DNA, so that the switch from low to high affinity for the operator is accomplished without a change of quaternary structure of the repressor (Fig. 50).

The reading heads change their tertiary structure in response to the binding of tryptophan (Fig. 51). Its indol ring binds in a hydrophobic pocket between the rigid core and helix E; its amino group forms hydrogen bonds with two main-chain carbonyls at the end of helix B, and it also compensates the partial negative charge of the helix dipole; its carboxyl group forms two hydrogen bonds with the guanidinium group of an arginine (84) and places that group in a position where it can form a hydrogen bond with one of the phosphates of the operator DNA. Replacement of arginine 84 by either cysteine or histidine inactivates the repressor (Kelly & Yanofsky, 1985). In the absence of tryptophan the pocket collapses and the guanidinium group of the same arginine forms hydrogen bonds with the main-chain carbonyls that were bound to the amino group of tryptophan. These bonds clamp the two reading heads too closely together to fit into successive grooves of the DNA double helix, while tryptophan moves them further apart and leaves them correctly spaced to do so. They also become more flexible.

How does the protein recognize the operator sequence? Rosenberg's structure of the *Eco*R1 restriction enzyme–DNA complex led us to expect interactions between amino acid side-chains and specific bases in the operator (Frederick *et al.* 1984), but in the *trp* repressor–operator complex most interactions are hydrogen bonds between the DNA phosphates and donor groups of the repressor, including one hydrogen bond with the indol imino group of the bound tryptophan (Fig. 52). The only hydrogen bonds between the repressor and specific bases are indirect and mediated by water molecules (Fig. 53). One links the OH_γ of a threonine (83) to an adenine (7) and two others link two main-chain imino groups and the amino group of a lysine (72) to another adenine (5) and a guanine (6). Symmetrical replacement of guanine 6 by any other base or of adenine 5 by a pyrimidine drastically reduces the affinity of the repressor for the operator *in vivo* (Bass *et al.* 1987). Repression is also diminished by replacing threonine 83 and lysine 72 by other residues. Replacement of glycine 78 by either aspartate or serine has similar effects (Kelly & Yanofsky, 1985); according to the structure there is not enough

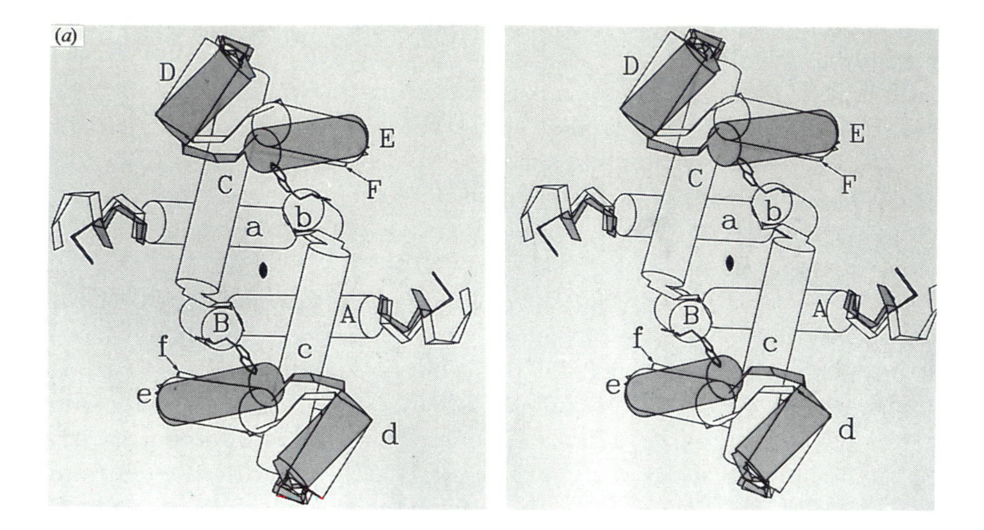

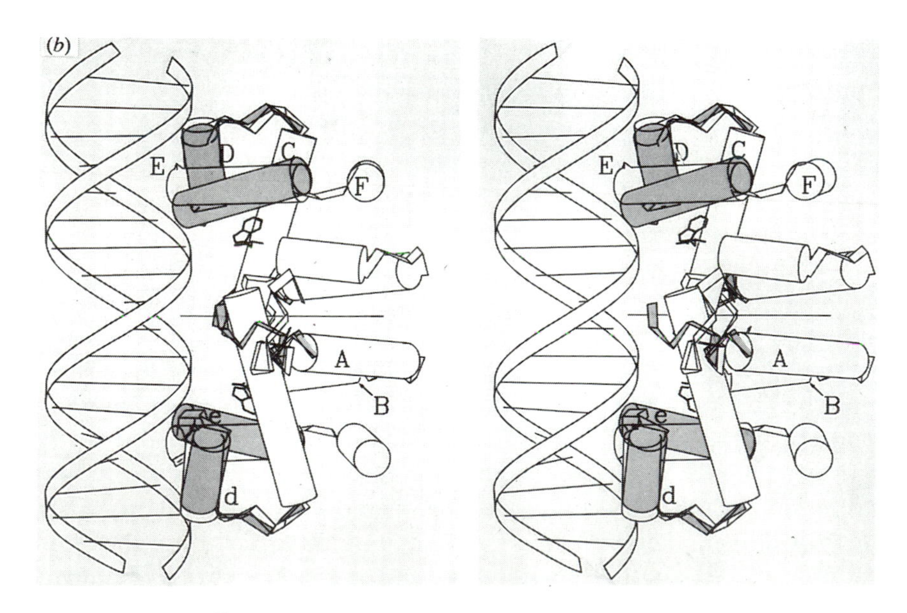

Fig. 50. Tertiary structure of the *trp* repressor dimer of *E. coli* with tryptophan (white) and without it (shaded). Cylinders represent α-helices, labelled A–E in one subunit and a–e in the other. In (*a*) the 'reading heads' connecting helices D and E, and d and e, face the observer; in (*b*) they face a double helix of DNA. Without tryptophan they are too close together to fit into successive major grooves. (From Zhang *et al.* 1987.)

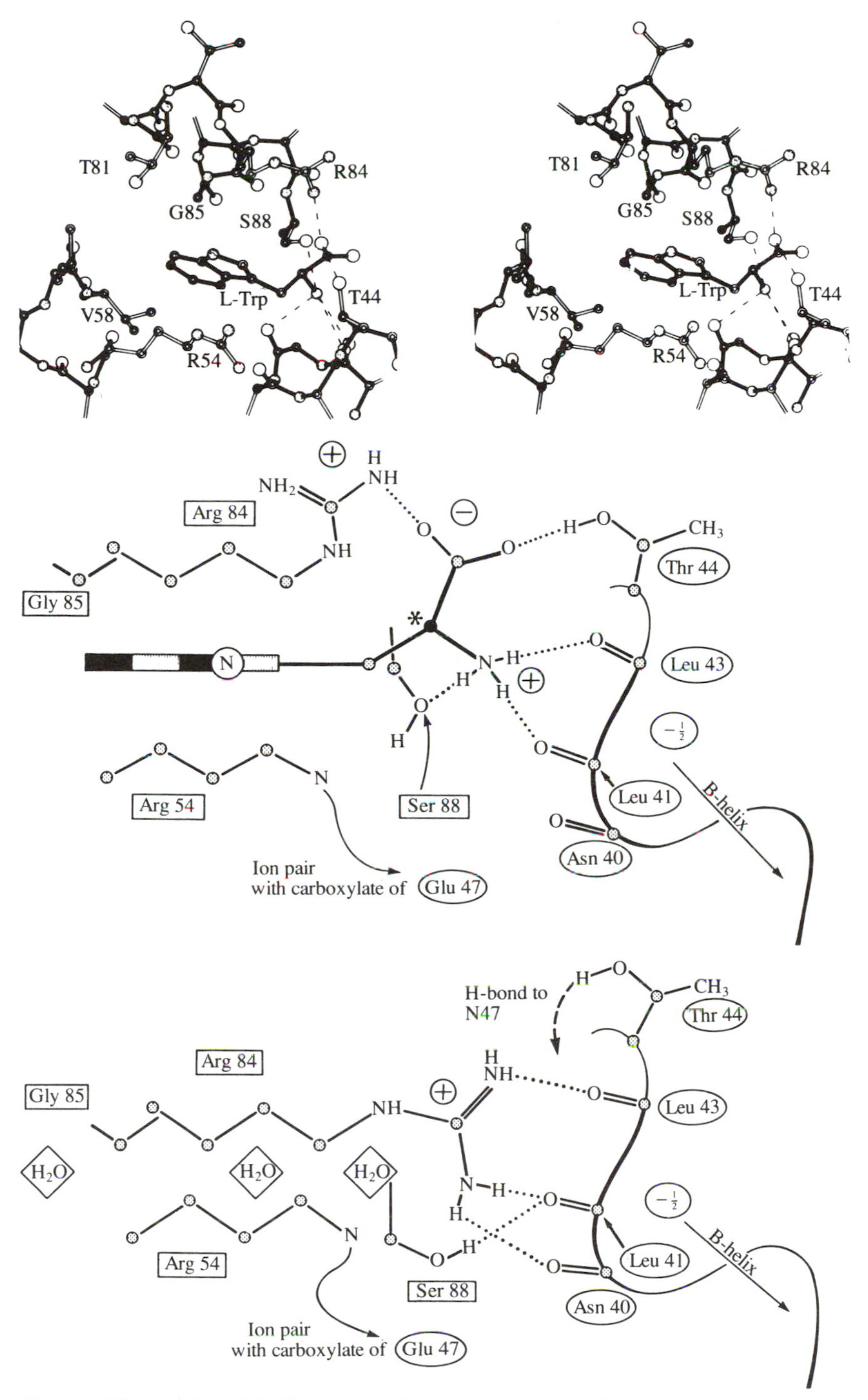

Fig. 51. Tryptophan binding sites of the *trp* repressor. The carboxylate of the tryptophan accepts a hydrogen bond from arginine 84 and threonine 44, and its amino group donates hydrogen bonds to the carbonyls of leucines 41 and 43. In the absence of tryptophan, the arginine donates hydrogen bonds to the same carbonyls instead. (*a*) shows a stereoview and (*b*) a diagrammatic view of the site with tryptophan, and (c) a diagrammatic view without it. ((*a*) Courtesy of Dr P. B. Sigler; (*b*) and (*c*) are from Zhang *et al.* (1987).)

room between the α-carbon of the glycine and the DNA to fit the side-chains in, quite apart from the repulsive effect of the aspartate. Both the reading heads of the repressor and the DNA bend on combining, whence Sigler *et al.* believe recognition of the operator sequence to depend partly on a specific distortion of the double helix allowed by the base sequence as well as on indirect hydrogen bonds with specific bases. In fact, if the dissociation constant of the repressor from the operator is about 10^3 times smaller than that from random DNA, only an extra 4000 calories are needed to account for that difference. There are six water-mediated hydrogen bonds linking the repressor to specific bases, whence only between 700 and 1000 calories per bond are required to provide that energy. For comparison, A. R. Fersht, G. Winter, D. M. Blow and their collaborators found that deletion of a hydrogen bond between an uncharged donor and acceptor in the active site of tyrosyl-tRNA synthetase weakened the binding energy between enzyme and substrate by 0·5–1·5 kcal/mol. Deletion of a hydrogen between a charged donor and uncharged acceptor or vice versa weakened the binding energy by a further 3 kcal/mol. This suggests that there may be no need to invoke any mechanism of discrimination between the operator DNA and random DNA other than the water-mediated hydrogen bonds to the operator-specific bases. (For review see Fersht *et al.* 1986.) The likelihood of the repressor being complementary to a bend of the operator DNA that is specific for its base sequence has been lessened by the discovery of a crystal structure of a dodecanucleotide in which neighbouring molecules are bent in opposite directions in response to lattice forces. This structure suggests that the double helix is bent easily in any direction, and that the bend seen in the trp repressor–operator complex is unlikely to have contributed significantly to the free energy of discrimination between operator and random DNA (Di Gabriele *et al.* 1989).

The electron density of the amino-terminal region is ill defined in crystals of both free and operator-bound repressor, suggesting that this region is mobile and does not contribute to the binding energy. In apparent contradiction, repressor from which the amino-terminal hexa-peptide with the sequence Ala-Gln-Gln-Ser-Pro-Tyr- had been cleaved, exhibited a 50-fold larger dissociation constant from both operator and random DNA (Carey, 1989). Did that peptide bind to the DNA in solution but not in the crystal? NMR studies showed it to be mobile also in solution (Arrowsmith *et al.* 1989). I do not think that the contradiction is real, because a factor of 50 amounts to only 1 kcal per repressor monomer, and this can easily be accounted for by non-specific water-mediated interaction between the α-NH_3^+ of alanine 1 and the DNA phosphates.

While tryptophan induces binding of the repressor to the operator, indolpropionic acid inhibits binding and therefore induces tryptophan synthesis. How can the loss of the amino group convert it from an inducer to an inhibitor? X-ray analysis shows that it inverts the orientation of the indol ring in the binding site: when tryptophan binds, its amino group interacts with the carbonyls at the end of helix B and its indol imino group forms a hydrogen bond with a phosphate oxygen of the DNA; when indolpropionate binds, the indol imino group makes the hydrogen bond to a carbonyl oxygen at the end of helix B, and the carboxylate

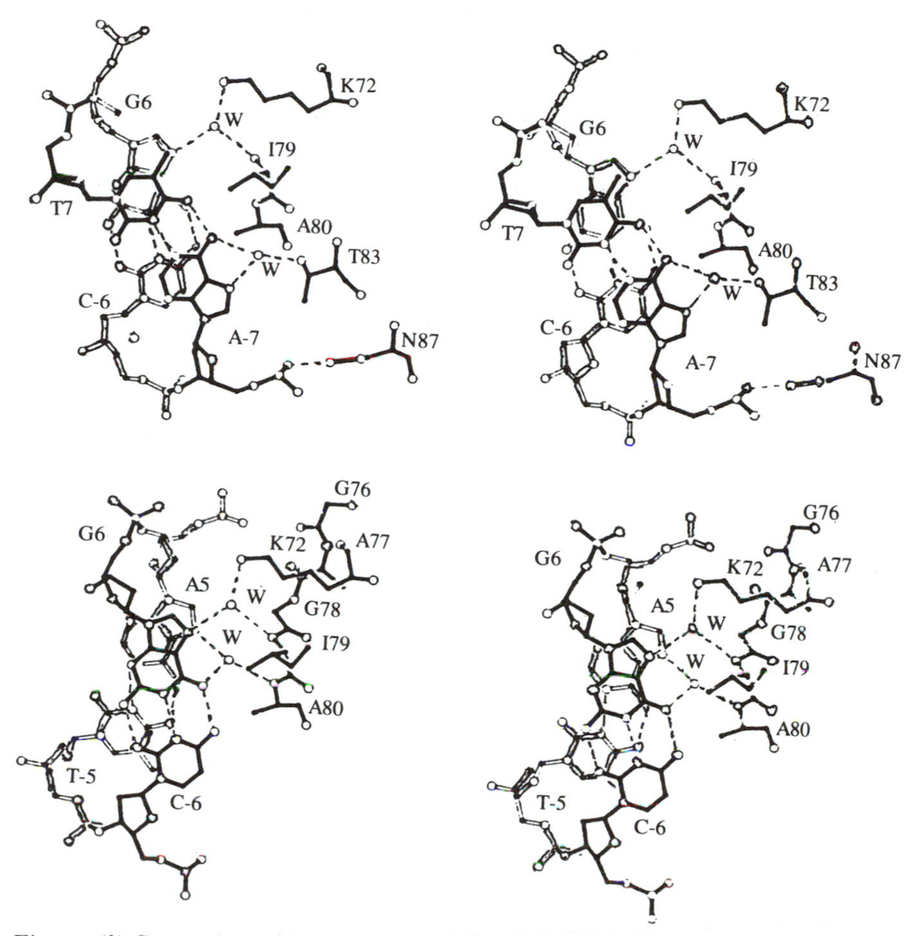

Fig. 53 (*b*) Stereoview of hydrogen bonds between the *trp* repressor and specific bases of the operator mediated by water molecules as described in the text. (From Otwinowski *et al.* 1988.)

overlaps with the position taken by the phosphate ester chain in the operator complex and repels the DNA (Fig. 54).

There is no direct evidence as yet that the binding of the two reading heads to the operator is cooperative, because it has not been possible to dissociate the two subunits and measure their individual dissociation constants from the operator, nor to measure the dissociation constant of half the palindromic operator sequence to the repressor, but the very small dissociation constant ($\sim 0{\cdot}2$–$0{\cdot}5$ nM) suggests cooperativity. How could this arise without a change of quaternary structure? Chemists have synthesized model compounds that exhibit cooperativity by purely entropic effects. Rebek and his colleagues have linked two cyclic ethers by phenyl groups, producing a twin compound capable of chelating two moles of $Hg(CN)_2$ (Figs. 55, 56). Its binding curve has a Hill's coefficient of 1·5, due to the association constant of $Hg(CN)_2$ with the second ring (K_2) being ten times larger than that with the first (K_1). The ratio of K1/K2 is independent of temperature. Since the enthalpies of binding are the same for both steps ($-7{\cdot}3 \pm 0{\cdot}3$ kcal mol^{-1}),

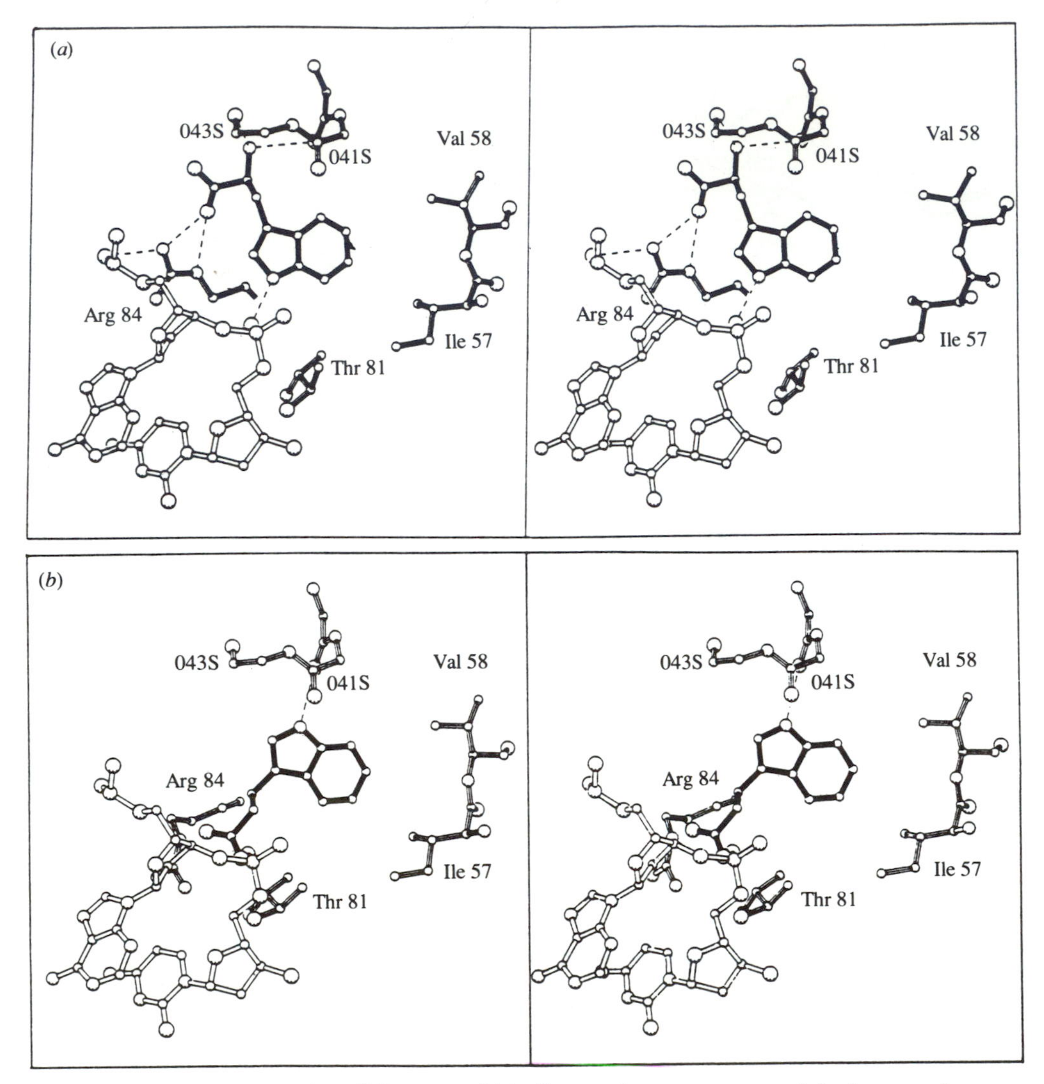

Fig. 54. Stereoviews showing difference of binding to the *trp* repressor (*a*) of tryptophan which represses the biosynthesis of tryptophan, and (*b*) of indolpropionic acid which induces it. Note the inversion of the indol ring and the overlap of the propionate with the phosphate chain in (*b*), while in (*a*) the indol NH donates a hydrogen bond to the phosphate oxygen (from Lawson & Sigler, 1988.)

the origin of cooperatively must be entropic. It arises because the metal-free rings are flexible. Binding of the first mercury-cyanide stiffens both macrocycles, so that the loss of entropy on binding the second mercury atom is smaller than the loss on binding the first (Rebek *et al.* 1981; Onan *et al.* 1983). Had the two mercury ions been linked by a flexible aliphatic chain, the difference between the losses of entropy would have been much greater, and K_2 would have been larger than K_1 by several orders of magnitude, because the second reaction would have become a monomolecular one. Such a model would have corresponded more closely to the *trp* repressor or to the IgG's discussed in the next section.

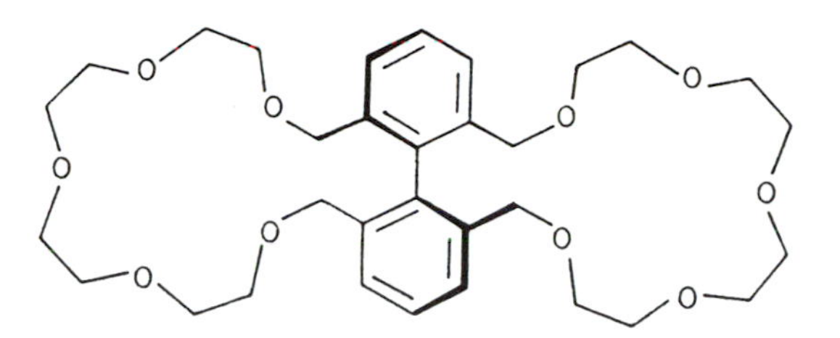

Fig. 55. Macrobicyclic ether. (From Rebek *et al.* 1981.)

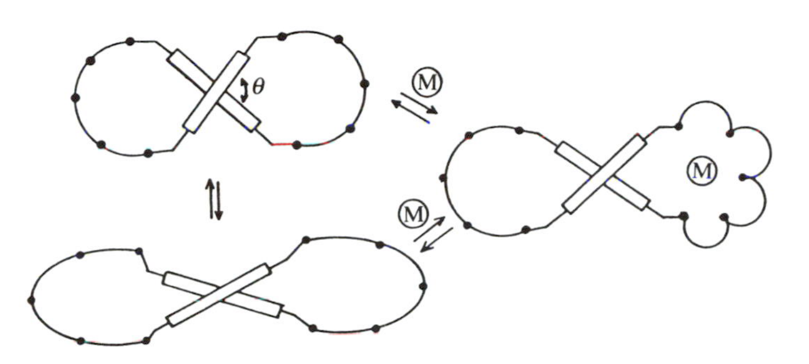

Fig. 56. Bottom left, loose conformation of metal-free ether; top left, conformation of ether ready for metal binding; right, binding of metal to one ring fixes the other in the binding conformation. (From Onan *et al.* 1983.)

In the tryptophan repressor, binding of the first reading head to the operator is paid for by losses in translational and rotational entropy of both the repressor and the operator. These losses are likely to be much smaller when the second reading head binds, because it will be already juxtaposed to the operator, resulting in cooperatively and a large reduction of the dissociation constant (Schleif, 1988).

In summary, cooperativity of binding the repressor to the operator, if it exists, works without any change in quaternary structure and may arise by entropic effects. Much larger cooperativity is likely to be manifested in the binding of the seven-fingered transcription factor TFIIIA to its tandem recognition sequence (Klug & Rhodes, 1987).

Like the *trp* repressor, the *met* repressor of *E. coli* was discovered by Cohen & Jacob (1959). In the presence of *S*-adenosylmethionine, a metabolite that is a vital donor of methyl groups, it represses transcription of the genes coding for the enzymes involved in the biosynthesis of methionine and *S*-adenosylmethionine; it also represses transcription of the gene coding for itself. The repressor is a dimer of two identical polypeptides of 104 amino acid residues and binds two molecules of *S*-adenosylmethionine.

Rafferty *et al.* (1989) and Phillips *et al.* (1989) have solved the structures of the aporepressor at 1·7 Å resolution and of its complex with *S*-adenosylmethionine at 1·9 Å resolution, sufficient to resolve all individual atoms other than hydrogen. The structure is unlike that of other repressors (Ptashne, 1986), because it contains no obvious helix-turn-helix motif. The two polypeptides are intertwined,

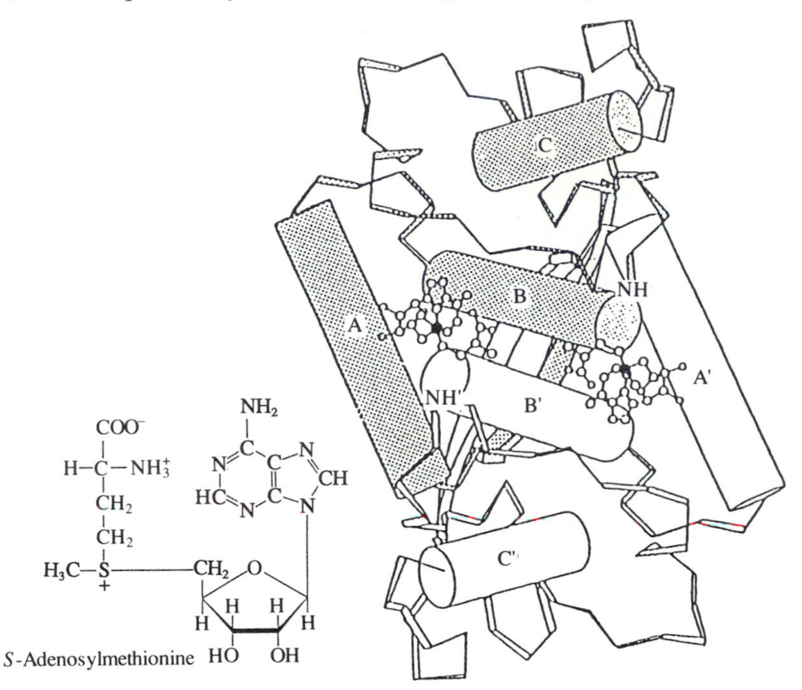

Fig. 57. The *met* repressor of *E. coli*. View parallel to the twofold symmetry axis, which is flanked by the two corepressors, seen from the direction of the DNA. The black circles mark the sulphur atoms. (Courtesy S. E. V. Phillips.)

and binding of the co-repressor induces no detectable changes in either quaternary or tertiary structure. The adenosine moiety of the co-repressor fits into a slot of the repressor which forms hydrogen bonds with the amino group of the methionine and with the adenine. The sulphur atom faces the surface (Fig. 57). Contact between the two monomers is made by the B-helices facing the reader and by the pair of antiparallel β-strands shown as partly hidden ribbons behind the B-helices. The structure of a double-helical 16mer polynucleotide with two tandem *met* operator consensus sequences linked to two S-adenosylmethionine *met* repressor dimers was recently solved by W. Somers and S. E. V. Phillips (1989). The antiparallel β-strands of the repressor fill the major groove of the DNA which closes up around them, causing the minor groove opposite to open up (Fig. 58). Eight hydrogen bonds link amino acid side chains to nucleotide bases. The S-adenosylmethionines face away from the DNA; unlike the tryptophan in the *trp* repressor, they play no direct part in operator binding. The 2-fold axis of the dimer coincides with that of the single 8mer operator and not, as preferred by Phillips *et al.* (1989), with the 2-fold axis relating the two operators.

The extensive contact along the A-helices joining the two repressor dimers appears to be essential for strong binding to the operators. In a filter-binding assay, half-maximal binding of the S-adenosylmethionine *met* repressor complex to the 16mer operator was observed at a repressor concentration of 45 nM. The binding curve was sigmoid, indicative of cooperative binding. A single 8mer

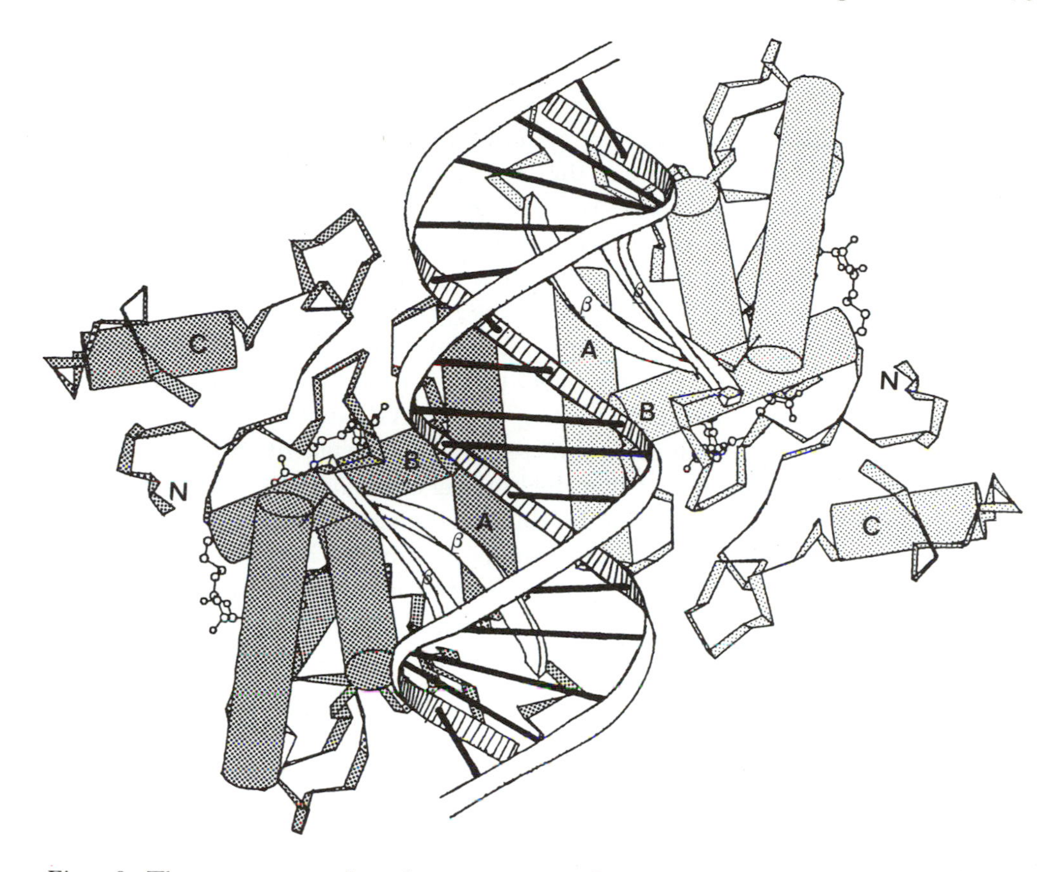

Fig. 58. The ternary complex of *met* repressor, S-adenosylmethionine and operator DNA (Courtesy J. B. Rafferty). The repressors wind around the DNA in a left-handed helix. One strand of the operator has the sequence 5′-TTAGACGTCTAGACGTCTT-3′; the sequence of the other strand is complementary.

operator sequence bound only weakly to the repressor complex; without S-adenosylmethionine, binding of the repressor to the 16mer operator sequence was non-cooperative and weak (Phillips *et al.* 1989). The structures solved so far do not reveal why S-adenosylmethionine raises the association constant of the repressor to the operator about 1·00-fold, because the structure of the aporepressor is indistinguishable from that of the S-adenosylmethionine repressor complex.

10. IMMUNOGLOBULINS: COOPERATIVE BINDING TO MULTIVALENT ANTIGENS

The binding of the tryptophan repressor to the operator is similar, kinetically and thermodynamically, to the binding of divalent antibodies (IgG's) to multivalent antigens. There are several reports of this being cooperative. For example, Gopalakrishnan & Karush (1974) immunized a rabbit with pneumococci conjugated wih p-aminophenyl-β-lactoside (PAPL) and isolated an electrophoretically homogeneous IgG fraction from it antiserum. They measured the binding of free *N*-acetyl-PAPL to this IgG by equilibrium dialysis and obtained a linear Scatchard plot giving an association constant of 2×10^5. They then made a multivalent antigen by reacting bacteriophage ϕX174 with *N*-bromoacetyl-PAPL and measured the rates of its neutralization by, and dissociation from, their IgG. The ratio of the two rate constants yielded an association constant of over 10^9, nearly 10000 times larger than that of free *N*-acetyl-PAPL to IgG. At first sight this means merely that the free energy of binding IgG to the virus was about twice as large as to the free hapten, as it would have to be, but bearing in mind that only a fraction of the IgG's that have bound to one hapten on the viral surface are likely to have found another hapten within reach, and that probably only a fraction of those IgG's that bound would have neutralized the virus, the enhancement factor is not inconsistent with cooperative binding. Karush (1978) points out that the binding of the first F_{ab} arm to the virus is a second-order reaction, while the binding of the second F_{ab} arm to the virus is a first-order reaction, and therefore entails no further loss of entropy. The enhancement factor for decavalent IgM antibodies is between a hundred and a thousand times greater still than for IgG's (Karush, 1978).

Crothers & Metzger (1972) had already worked out a theory for antibody binding based on statistical arguments. Once the first F_{ab} segment has bound to a multivalent antigen particle such as a virus, the effective concentration of the second F_{ab} segment becomes much greater than the real IgG concentration, because it is confined within a small volume determined by the flexibility of the hinge between the two F_{ab}'s. If K_1' is the association constant of one F_{ab} to a particle of area a and antigen determinant density x, then K_2', the association constant of the second F_{ab} to the same particle, is defined by

$$K_2' = 3K_1'(V/2\langle r\rangle Na).$$

N is the number of antigenic particles in volume V in the standard state used to define K_1' (say in 1 mol/l) and $\langle r\rangle$ is the distance between the two F_{ab}'s averaged over all angles of the hinge between them. If we call the expression in parentheses the cooperativity factor, we see that it is inversely proportional to $\langle r\rangle$. By the law of mass action the association constant K_{obs}' of a bivalent antibody to a multivalent determinant is

$$K_{obs}' = 2K_1'(1+K_2').$$

The factor 2 arises from degeneracy through interchange between the two binding sites. When $K_2' \geqslant 1$, $K_{obs}' = 2K_2'$.

Crother & Metzger's statistical treatment is equivalent to Rebek *et al.*'s thermodynamic analysis of cooperative metal binding to the macrobicyclic ethers described in the preceding section. Once the first F_{ab} has bound, most of the loss of rotational and translational entropy has been paid for; hence the second F_{ab} binds with higher affinity. The proposed dominance of the entropic factor could

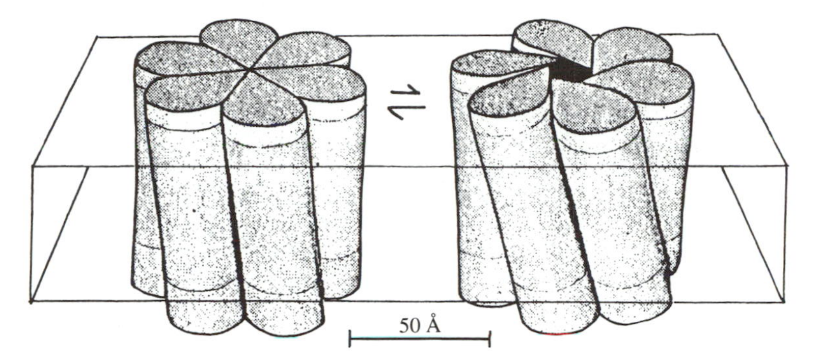

Fig. 59. Diagrammatic representation of gap junction from rat liver in the closed and open conformations, suggested by X-ray and electron microscope studies. The subunits turn about rotation axes that are normal to the central symmetry axis as in Figure 2, p. 3. The rotation axes are at the bottom of the diagram; concerted rotation of the subunits opens or closes the pore at the top, cytoplasmic end of the channel. (From Unwin, 1987.)

be tested by carrying out Gopalakrishnan & Karush's experiment at different temperatures.

II. ALLOSTERIC MEMBRANE PROTEINS

In eukaryotic cells most control and all communication relies on membrane proteins which respond to chemical or electrical signals. No detailed X-ray structures have yet been obtained of any such proteins, but cryoelectron microscopy has provided first glimpses of the allosteric changes undergone by two of them: the gap junction and the nicotinic acetylcholine receptor (Unwin, 1987; Changeux *et al.* 1987).

Gap junctions are channels of communication from cell to cell that close in response to H^+ and Ca^{2+}. They consist of six identical subunits of molecular weight 32000, each containing four stretches of 20 mainly hydrophobic amino acid residues, reminiscent of the composition of the transmembrane α-helices in the purple membrane protein and the photochemical reaction centre of purple bacteria. The structure of crystalline double layers of gap junctions from rat liver has been analysed by a combination of X-ray diffraction, negative-staining electron microscopy and cryoelectron microscopy of unstained specimens, followed by three-dimensional image reconstruction. The images show six particles 70–80 Å long and about 30 Å wide, grouped symmetrically around a central channel. The channel is about 25 Å wide at the contact between its opposite from the neighbouring cell and narrows towards its opening on the cytoplasmic side. Specimens frozen from buffers with and without 50 μM $CaCl_2$ look distinctly different. Without calcium the rods are tilted to the channel axis by 17°; addition of calcium makes them stand more upright by 7·5° (Fig. 59). They turn about axes of rotation that are normal to the channel axis and intersect it at the extracellular end, so that the change of tilt could alter the width of the channel at the cytoplasmic end by $2 \times 70 \sin 7{\cdot}5 = 18$ Å (Unwin & Zampighi, 1980; Unwin

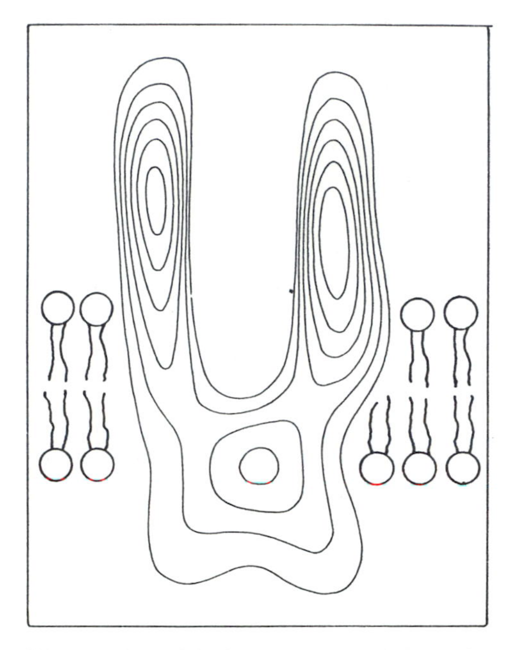

Fig. 60. Acetylcholine receptor of *Torpedo marmorata* in cross-section. Only the wide extracellular portion of the ion channel is resolved; the allosterically operated, narrower channel in the cytoplasmic region appears as continuous density. (From Unwin, 1987.)

& Ennis, 1984). In summary, the gap junction is a hexameric protein that undergoes a symmetrical change in quaternary structure of the kind shown in Fig. 2 p. 3. The symmetry implies that the change should be concerted and cooperative with respect to [Ca^{2+}], though this has not yet been tested, nor have the calcium binding sites been located, though the typical calcium binding fold might reveal itself by sequence homology with that of other calcium-binding proteins.

The acetylcholine receptor is called *nicotinic* because nicotine inhibits the depolarization of muscle cells caused by acetylcholine following its release at the neuromuscular junction. (Other acetylcholine receptors are blocked by muscarine instead.) The receptor consists of four distinct, but markedly homologous polypeptides and has the composition $\alpha_2\beta\gamma\delta$. Each of these polypeptides contains four or five stretches of 20 mainly hydrophobic residues indicative of membrane-spanning α-helices. The five subunits form a hollow cylinder that is about 70 Å wide and 140 Å long, and protrudes from both sides of the synaptic membrane. Its central channel is about 25 Å wide at its presynaptic entry and narrows at its postsynaptic exit (Fig. 60; see Toyoshima & Unwin, 1988). In its resting state, the receptor is impermeable to ions. In response to a 1 ms pulse of acetylcholine it becomes permeable to Na^+ and K^+ for about 3 ms. On prolonged exposure to acetylcholine it remains closed, regardless of further pulses of the transmitter. This state is called 'desensitized'.

The receptors in the electric organ of *Torpedo* can be extracted, purified and crystallized in the form of long tubes. Unwin *et al.* (1988) examined such crystals in buffer with and without carbamylcholine, a stable analogue of acetylcholine, as a desensitizer, using cryoelectron microscopy at different angles of tilt and three-dimensional image reconstruction. Statistical analysis of the images showed significant differences between the two preparations. In response to carbamylcholine the γ-subunits bulged outwards and parts of the δ-subunits turned about an axis perpendicular to the central pseudo-fivefold axis of symmetry (Fig. 61). In other words, the γ-subunits perform one, and parts of the δ-subunit another of the movements expected in an allosteric protein with subunits surrounding a central symmetry axis. More detailed analysis of the allosteric change awaits the preparation of ordered three-dimensional crystals of the receptor.

Several experiments have shown the response of the receptor to transmitters to be cooperative. Cash & Hess (1980) made membrane vesicles rich in acetylcholine receptors from the electric organ of *Electrophorus electricus* and measured the flux of $^{86}Rb^{+}$ across the membrane as a function of the concentration of carbamylcholine. The response curve of ion flux to ligand was sigmoidal and fitted a mechanism requiring two molecules of carbamylcholine to open one receptor channel by altering the allosteric equilibrium between an active, open and an inactive, closed state. Trautmann (1984) determined the dose–response curve of frog muscle fibres by application of voltage jumps and comparison of the resulting ion current in the absence of an agonist and in the presence of varying concentrations of the agonist carbamylcholine. Subtraction of the former current from the latter provided a measure of the number of ion channels that opened in response to the agonist. A double logarithmic plot of current versus [carbamylcholine] gave a straight line with a slope of 2, indicative of cooperative binding to two identical sites. Some at least of the residues contributing to these sites have been located on the presynaptic faces of the two α-subunits (Fig. 62).

Various compounds inhibit ion flow through the channel even though they do not react with the acetylcholine-binding sites. They are called non-competitive blockers. The binding sites for one such blocker, chlorpromazine, has been located by photo affinity labelling of serines 254β and 262δ (Giraudat *et al.* 1987). These serines are part of the membrane-spanning channels and are about 40 Å from the acetylcholine binding sites. The location of serines in the channel is intriguing. $O_{\gamma}H$ of serine in an α-helix donates a hydrogen bond to the carbonyl oxygen in the next helical turn, so that its lone-pair electrons point away from the helix. A ring of serines from the α-helices facing the channel might therefore act as a cation-binding macrocycle.

Acetylcholine acclerates the binding of chlorpromazine to these serine sites. Using postsynaptic membrane fragments from *Torpedo* that are rich in receptor, Changeux and his collaborators have measured the kinetics of chlorpromazine binding as a function of [acetylcholine] and found evidence for at least three major conformations (Heidmann *et al.* 1983). Their dissociation constants vary from 3–5 μM in resting-active 'open' receptors to 1 μM in the fast-desensitized intermediate state, to 3–5 nM in the slowly desensitized state. In addition, there is the resting-inactive 'closed' state.

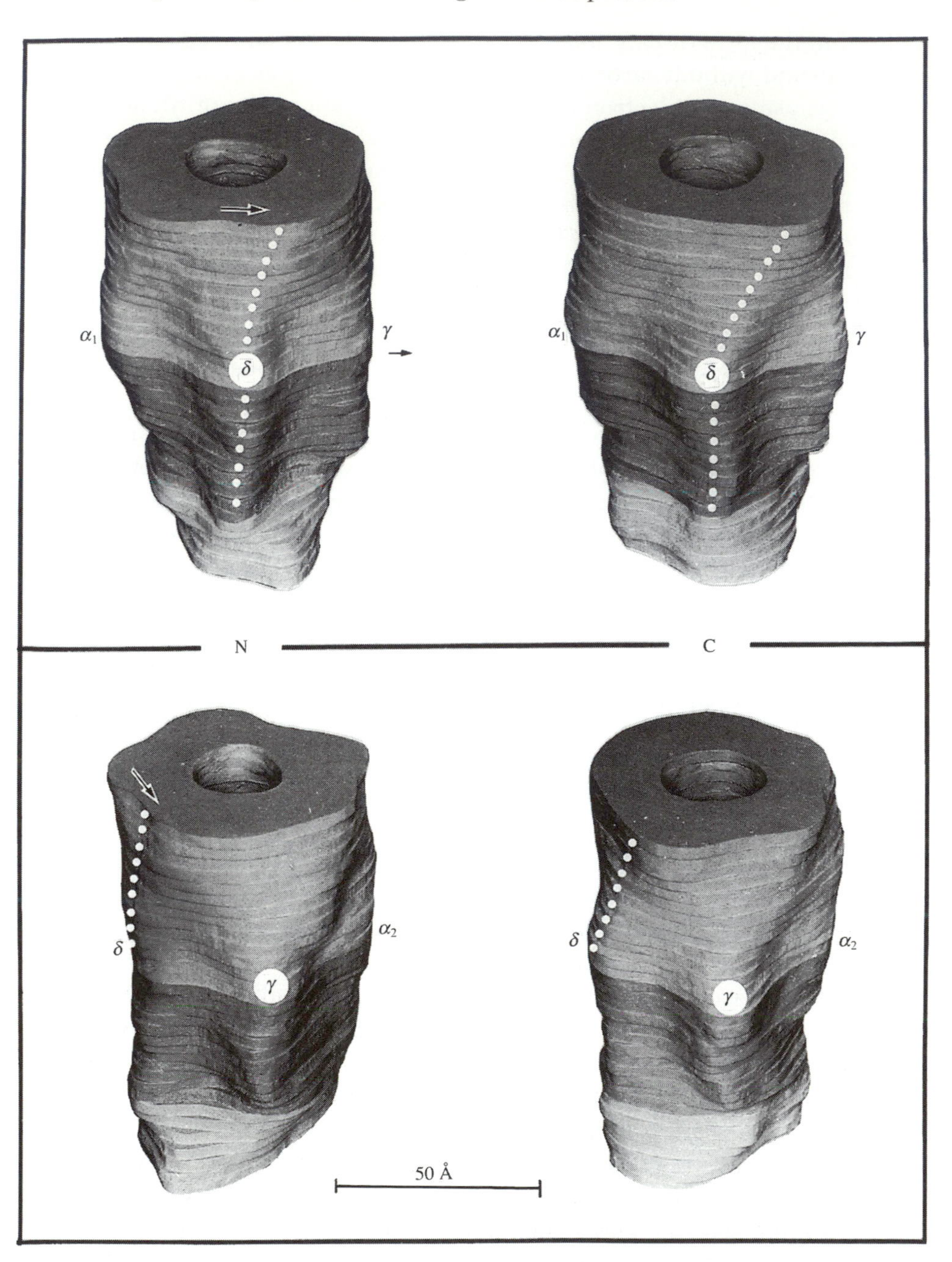

Fig. 61. Three-dimensional reconstruction of the acetylcholine receptor, before (left) and right after desensitization by carbamylcholine, looking towards the δ-subunits in the upper pair and the γ-subunits in the lower pair. On exposure to carbamylcholine, the δ-subunit tilts and the γ-subunit bulges outwards. (From Unwin *et al.* 1988.)

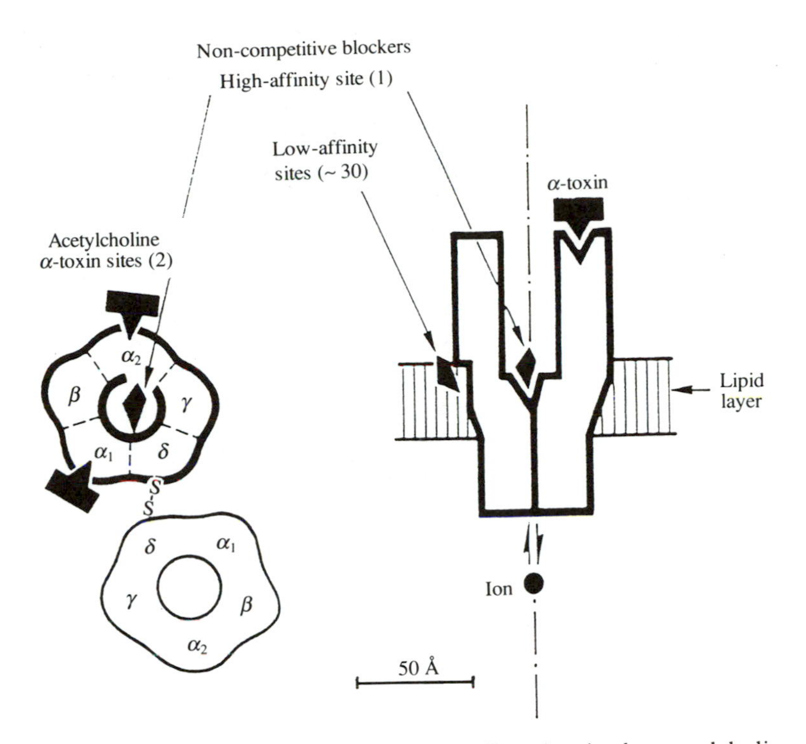

Fig. 62. Disposition of subunits and binding sites in the acetylcholine receptor. α-toxin blocks the acetylcholine-binding site. Chlorpromazine binds to high-affinity sites in the ion channel. The sites for non-competitive blockers of low affinity are not known in detail. (From Changeux *et al.* 1987.)

The role of the serines has been corroborated by directed mutagenesis. mRNAs transcribed *in vitro* from cDNA clones of mouse acetylcholine receptor subunits were injected into *Xenopus* oocytes, where receptor was synthesized and incorporated into the cell membrane. Such oocytes can serve as model receptor membranes. The serines of the δ- and the two α-chains were replaced by alanines. The mutant receptors displayed abnormally low outward conductance, while their other properties were normal. A tertiary ammonium derivative of lidocaine entered ion channels while they were open and blocked them temporarily, presumably bound to the chelating groups in the channel. Replacement of one, two or three of the serines by alanines progressively shortened the lifetime of the blocked state, while replacement of a phenylalanine in the channel by a serine lengthened it (Leonard *et al.* 1988), consistent with binding of the cation by the serines.

Oiki *et al.* (1988) synthesized a peptide having the same amino acid sequence as the transmembrane helix of the δ-subunit that carries the chlorpromazine-labelled serine. A lipid bilayer incorporating these 23-residue peptides mimicked the ion transport behaviour of the acetylcholine receptor. The peptides formed channels that alternated between open and closed states and exhibited conductance selectivity for Na^+, K^+ and $Tris^+$ in the ratio of 1.0:1.2:0.18. Atomic models of five such α-helical peptides were assembled around a central axis, with their hydrophilic residues inside and hydrophobic outside. They formed a channel of 4–6 Å diameter, wide enough for Na^+ and K^+ and tight for $Tris^+$.

Lear *et al.* (1988) have mimicked the conductance of the acetylcholine receptor by incorporating a much simpler synthetic *a*-helical peptide with the constitution H_2N-(Leu-Ser-Ser-Leu-Leu-Ser-Leu)$_3$-$CONH_2$ into a phospholipid bilayer, thus demonstrating once more the crucial role of the serines.

Carmichael *et al.* (1989) have tried to mimic the constellation of serines in biological ion channels by synthesizing a crown ether with six oxygens in a ring and attaching to it hydrophobic side chains that would tend to form a channel wall when immersed in lipid bilayer vesicles. They imposed a pH gradient of pH 6·5 inside and pH 5·5 outside the vesicles. In the absence of their model channel, this gradient collapsed slowly over several days. In the presence of 0·5 nM of their channel, it collapsed in less than 3 minutes. Gramicidin D, a known channel-forming antibiotic, made it collapse in one minute. The authors stress that the precise mode of action of their artificial ion channel has yet to be established, but their success makes it clear that highly active ion transporters can be created synthetically by modelling those found in nature. Jullien & Lehn (1988) have synthesized a macrocycle containing six ether oxygens to which they attached eight long polyether chains, four on each side, to mimic the hydrophobic channel of biological ion transporters, but they have not yet studied its function.

R. Lutter and J. E. Walker in the MRC Laboratory of Molecular Biology have grown three-dimensional crystals of the *mitochondrial ATPase* from beef heart, an allosteric protein that also acts as a proton pump. The best of them diffract X-rays to a spacing of about 3 Å, which should be sufficient for locating the course of the polypeptide chain and the orientation of most side-chains (J. E. Walker, private communication). Structures of membrane proteins to at least that resolution will be needed if we are to understand the action of transmitters, hormones, second messengers and growth factors. Most drugs act on membrane proteins, and drug design will have to make do with indirect analyses of binding sites as long as their detailed structures remain unknown. This is the greatest challenge for protein crystallography in the future.

12. DISCUSSION

Most of Monod *et al.*'s (1965) predictions have been brilliantly borne out by the structural analyses described here. The subunit contacts of haemoglobin, phosphofructokinase and aspartate transcarbamylase are very clearly designed to allow only concerted transitions between alternative quaternary structures. The alternative contacts are stabilized predominantly by electrostatic interactions that take the form of hydrogen bonds, mostly between side-chains of opposite charge. We may define relaxed quaternary structures as those whose activity is the same as that of the free subunits, and tense ones as those where additional bonds between the subunits either lower or raise their activity. According to this definition human deoxyhaemoglobin and deoxyhaemocyanin of the spiny lobster have the tense structure, oxyhaemoglobin and oxyhaemocyanin the relaxed one. Inactive aspartate transcarbamylase and phosphorylase *a* have the tense structure, active aspartate transcarbamylase and phosphorylase *b* the relaxed one. Since isolated dimers of phosphofructokinase lack catalytic sites, the epithets 'tense' and 'relaxed' are inappropriate for its two quarternary structures, neither of which is constrained by additional bonds between the subunits.

The active sites of haemoglobin and haemocyanin are buried in folds of the polypeptide chain; in haemoglobin the major fold is coded for by a single exon, allowing activity to have evolved in the monomer. In the allosteric enzymes, on the other hand, the active sites lie between domains or between subunits, and evolution of catalytic activity would have required complementation by the products of more than one exon, perhaps by gene duplication. Catalytic sites at the boundaries between domains or subunits may have favoured the evolution of regulatory controls, since it is easier to change catalytic activity by shifting them relative to each other than by altering the internal structure of domains. On the other hand, regulation of oxygen affinity in haemoglobin and haemocyanin merely needed simple push–pull devices.

Two implications of Monod, Wyman and Changeux's theory have not been borne out by experiment, at least not in haemoglobin, the only protein where they have been tested rigorously.

'Note that by reason of symmetry and because the binding of any one ligand molecule is assumed to be intrinsically independent of the binding of any other, the microscopic dissociation constants are the same in all homologous sites of the two states.

'According to the model, heterotropic effects should be due exclusively to the displacement of the spontaneous equilibrium between the R and T states of the protein.'

The number of protons released by a haemoglobin solution rises linearly with the number of oxygen molecules taken up. Initially, these protons come from the rupture of salt bridges in the T-structure, which raises its oxygen affinity. Hence uptake of oxygen by any one subunit in the T-structure raises the oxygen affinity of the next one, which makes the mechanism sequential rather than concerted, consistent with the theory of Koshland *et al.* (1966). Furthermore, heterotropic ligands, mutations or chemical modifications that modify the allosteric constant L always affect K_T, because any bond that influences the allosteric equilibrium also contributes to the tension at the haems.

Aspartate transcarbamylase and phosphofructokinase do not exhibit quite the same behaviour because the catalytic activity of their T-structures is negligible, but it seems that any changes in L also affect K_M and V_{max} of their R-structures. Most proteins are held together by weak forces, whence any local steric or electrostatic changes are liable to make themselves felt throughout. In phosphorylase both the *a* and *b* forms are subject to activation and inhibition by varying degrees, and each can exist in either the R or the T structure; sequential and concerted behaviour may exist side by side.

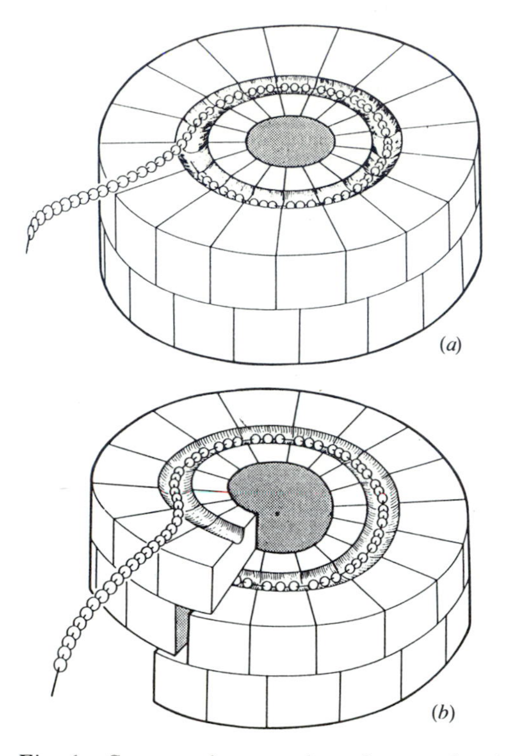

Fig. 63. Suggested conversion of a 17-subunit disk of tobacco mosaic virus protein to a helix by combination with the helix-initiating loop of the viral RNA. (Courtesy A. Klug.)

Monod *et al.* (1963) expected the tryptophan repressor to be an allosteric protein, but its behaviour does not accord with the definitions formulated by Monod *et al.* (1965), because the structural changes induced by tryptophan are not concerted. If binding to the operator is cooperative, as seems likely, then the cooperativity is due to simple entropic effects, as in the binding of bivalent antibodies to multivalent antigens. The surface area of one of the reading heads of the repressor buried in its complex with the operator is about the same as that of the F_{ab} fragment buried in its complex with lysozyme: 600 Å^2 (Amit *et al.* 1986). In fact, the activated tryptophan repressor with its two flexible reading heads behaves like an IgG antibody to the operator rather than an allosteric protein. By contrast, the met repressor is activated without any changes in quaternary or tertiary structure; the co-repressor itself seems to supply the additional bond energy needed for its binding to the operator.

Concerted allosteric transitions may play a part in macromolecular assembly. Such a transition has been found when tobacco mosaic virus was reassembled *in vitro* from its component protein molecules and RNA. The virus is a long rod made up of a single-stranded, helical chain of RNA surrounded by a helical array of small protein molecules. There are $16\frac{1}{3}$ of these per turn. Reassembly was found to be fastest when viral RNA was mixed with a suspension of protein aggregates, consisting mainly of double disks, each made up of 17 protein molecules in a ring.

The rings are threaded on to the RNA and then undergo a transition to the helical structure found in the virus (Fig. 63). In the absence of RNA the transition from disk to helix can be triggered by lowering the pH of a suspension of disks from 7 to 5, because helix formation is accompanied by the uptake of protons (Butler & Durham, 1977; Butler & Klug, 1978; Bloomer & Butler, 1986).

The structures of both the disk and the complete virus are now known at near-atomic resolution (Bloomer *et al.* 1978; Namba *et al.* 1989). Comparison shows that the transition from ring to helix involves changes in the contacts between subunits within the disk, accomplished largely by altered conformations of side-chains, and also drastic changes in the contacts between successive layers of subunits. In that process hydrogen bonds are formed between a glutamate in one layer and aspartates in the neighbouring layer; one carboxylate has its pK_a lowered by neighbouring arginines and the other has its pK_a raised by proximity of its neighbour's negative charge. Formation of that carboxylate pair must be responsible for the uptake of protons in the transition from disk to helix. The transition is accompanied by major changes in tertiary structure of the individual subunits; all the indications are that these are concerted. Tobacco mosaic virus is unlikely to be the only instance where concerted allosteric transitions play a part in morphogenesis.

pH-dependent allosteric transitions have also been observed in spherical plant viruses. Robinson & Harrison (1982) have made an X-ray study of such a transition in Tomato Bushy Stunt Virus. Its protein shell has the point group symmetry 532, with 180 identical subunits of MW 40000 forming a pattern of alternate hexagons and pentagons. Contacts between the subunits include carboxylates coordinated to calcium ions. If these are leached out and the pH is raised, repulsion between the carboxylates forces the subunits apart and swells the virus. That swelling is accomplished by the subunits' concerted rotations about and translations along radially directed axes. The rotations are small (2–4·5°), and the translations very large (18–19 Å). The allosteric transition may play a part in either the assembly or disassembly of the virus, or it may be no more than a laboratory curiosity.

Koshland (1959) predicted that certain enzymes may become catalytically active only after a change of structure induced by the binding of substrates. He called this *induced fit*. His theory has been verified by X-ray analysis in yeast hexokinase (Bennett & Steitz, 1978) and in citrate synthase (Remington *et al.* 1982). Hexokinase is monomeric or dimeric depending on the solvent; citrate synthase is always dimeric. The active sites of these enzymes lie in surface clefts that close up like nutcrackers on binding substrates, but there is no evidence of cooperativity between the two active sites of the dimers nor of feedback inhibition at sites that are separate from the catalytic ones. These enzymes are therefore outside the scope of my review, but the stereochemical mechanisms that operate the opening and closing of their active sites are similar to those found in allosteric transitions.

Chothia & Lesk (1985) have shown that the structural transitions of hexokinase and citrate synthase are achieved by small shifts of α-helices in the domains flanking the active sites. α-helices can shift relative to each other by up to 1·6 Å and turn by several degrees with only minor adjustments of their side chains, and such movements can be amplified by leverage when they are coupled in series. For example, in citrate synthase an initial shift by 0·9 Å and a rotation by 7° of one helix relative to the enzyme's rigid core is amplified to a shift by 10 Å and a rotation by 28° three helices away. Figure 50, p. 68, shows how the binding of tryptophan to the *trp* repressor rearranges its α-helices. Chothia & Lesk found that movements of α-helices also facilitate the closure of the active site in the T → R transition of aspartate transcarbamoylase. In phosphorylase, shifts of α-helices triggered by the phosphorylation of serine 14 seem to set in motion the change in quaternary structure that results in the rearrangement of the tower helices shown in Fig. 28, p. 39. On oxygenation of haemoglobin, helices F shift by 1·3–1·6 A and helices E by 0·7–1·1 A relative to the rigid BGH core. Their motion triggers the T → R switch along the $\alpha_\alpha\beta_2$ and $\alpha_2\beta_1$ contacts shown in Fig. 17, p. 20. Flexibility of helix packing appears to be essential to the function of many problems.

13. ACKNOWLEDGEMENTS

I thank Dr Arthur Lesk for generating Fig. 44, 46 and 47 on our computer graphics, Dr R. J. Fletterick for Fig. 25 and 27, Dr D. Barford for Fig. 28, Dr Louise Johnson for the diagrams on p. 34 and p. 42, Dr W. N. Lipscomb for Fig. 41 and 45, Dr David Eisenberg for Fig. 49, Drs Tilman Schirmer and P. R. Evans for Fig. 32–38, Dr S. E. V. Phillips for Fig. 57 and 58, and several colleagues for allowing me to see and describe their results before publication. I am grateful for support from the Medical Research Council, National Science Foundation grant no. PCM-8312414 and National Institutes of Health grant no. HL-31461.

14. REFERENCES

ABELSON, P. H, (1954). Amino acid biosynthesis in *Escherichia coli*: isotopic competition with ^{14}C glucose. *J. biol. Chem.* **206**, 335–343.

ALMASSY, R. J., JANSON, C. A., HAMLIN, R., XUONG, N.-H. & EISENBERG, D. (1986). Novel subunit–subunit interactions in the structure of glutamine synthetase. *Nature* **323**, 304–309.

AMIT, A. G., MARIUZZA, R. A., PHILLIPS, S. E. V. & POLJAK, R. J. (1986). Three-dimensional structure of an antigen–antibody complex at 2·8 Å resolution. *Science* **233**, 747–753.

ANTONINI, E. & BRUNORI, M. (1971). *Hemoglobin and Myoglobin and their Reactions with Oxygen*, pp. 241, 394. Amsterdam: North Holland.

ANGEL, W.-L., KARPLUS, M., POYART, C. & BURSEAUX, E. (1988). Analysis of proton release in oxygen binding by haemoglobin: implications for the cooperative mechanism. *Biochemistry* **27**, 1285–1301.

ARMSTRONG, W. H. & LIPPARD, S. J. (1984). Reversible protonation of the oxo bridge in a haemerythrin model compound. *J. Am. Chem. Soc.* **106**, 4632–4633.

ARMSTRONG, W. H., SPOOL, A., PAPAEFTHYMIOU, G. C., FRANKEL, R. B. & LIPPARD,

S. J. (1984). Assembly and characterisation of an accurate model for the diiron center in hemerythrin. *J. Am. Chem. Soc.* **106**, 3653–3667.

Arnone, A. (1972). X-ray diffraction study of binding of 2,3-diphosphoglycerate to human deoxyhaemoglobin. *Nature* **237**, 146–149.

Arrowsmith, C. H., Carey, J., Treat-Clemons, L. & Jardetzky, O. (1989). NMR assignments for the amino-terminal residues of trp repressor and their role in DNA binding. *Biochemistry* **28**, 3875–3885.

Baldwin, J. M. (1975). Structure and function of haemoglobin. *Progr. Biophys. molec. Biol.* **29**, 225–320.

Baldwin, J. M. & Chothia, C. (1979). Haemoglobin: the structural changes related to ligand binding and its allosteric mechanism. *J. molec. Biol.* **129**, 183–191.

Barford, D. & Johnson, L. N. (1989). The allosteric transition of glycogen phosphorylase. *Nature* **340**, 609–616.

Bass, S., Sugiono, P., Arvidson, D. N., Gunsalus, R. P. & Youderian, P. (1987). RNA specificity determinants of *E. coli* tryptophan repressor binding. *Genes & Development* **1**, 565–572.

Bennett, W. S. & Steitz, T. A. (1978). Glucose-induced conformational change in yeast hexokinase. *Proc. natn. Acad. Sci. U.S.A.* **75**, 4848–4852.

Berger, S. A. & Evans, P. R. (1990). Active site mutants altering the cooperativity of *E. coli* phosphofructokinase. *Nature* (in the Press).

Blackburn, M. N. & Schachman, H. K. (1977). Allosteric regulation of aspartate transcarbamylase. Effect of active site ligands on the reactivity of sulfydryl groups of the regulatory subunits. *Biochemistry* **16**, 5084–5090.

Blangy, D., Buc, H. & Monod, J. (1968). Kinetics of the allosteric interactions of phosphofructokinase from *Escherichia coli*. *J. molec. Biol.* **31**, 13–35.

Bloomer, A. C. & Butler, P. J. G. (1986). Tobacco mosaic virus: structure and assembly. In *The Plant Viruses* (ed. M. B. V. van Regenmortel and H. Fraenkel-Conrat). Plenum.

Bloomer, A. C., Champness, J. N., Bricogne, G., Staden, R. & Klug, A. (1978). Protein disk of tobacco mosaic virus at 2·8 Å resolution showing interactions within and between the subunits. *Nature* **276**, 362–368.

Bolognesi, M., Cannillo, E., Ascenzi, P., Giacometti, G. M., Merli, A. & Brunori, M. (1982). Reactivity of ferric *Aplysia* and sperm whale myoglobins towards imidazole: X-ray and binding study. *J. molec. Biol.* **158**, 305–315.

Bonaventura, C., Sullivan, B., Bonaventura, J. & Bourne, S. (1974). CO binding by hemocyanins of *Limulus polyphemus*, *Busycon carica* and *Callinectes sapidus*. *Biochemistry* **13**, 4784–4789.

Briehl, R. W. (1963). The relation between the oxygen equilibrium and aggregation of subunits in lamprey hemoglobin. *J. biol. Chem.* **238**, 2361–2366.

Brown, J. M., Powers, L., Kinkaid, B., Larrobee, J. A. & Spiro, T. G. (1980). Hemocyanin active site characterization and resonance Raman spectroscopy. *J. Am. Chem. Soc.* **102**, 4210–4216.

Brunori, M., Kuiper, H. A. & Zolla, L. (1982). Ligand binding and stereochemical effects in hemocyanins. *EMBO J.* **1**, 329–331.

Bunn, H. F. & Forget, G. B. (1985). *Hemoglobin, Molecular and Clinical Aspects*. Philadelphia: W. B. Saunders.

Butler, P. J. G. & Durham, A. C. H. (1977). Tobacco mosaic virus protein aggregation and the virus assembly. *Adv. Prot. Chem.* **31**, 187–251.

Butler, P. J. G. & Klug, A. (1978). The assembly of a virus. *Sci. Am.* **239**, 52–59.

Calhoun, D. B., Vanderkooi, J. M., Woodrow, G. V. III & Englander, S. W. (1983). Penetration of dioxygen into proteins studied by quenching of phosphorescence and fluorescence. *Biochemistry* **22**, 1526–1532.

Carey, J. (1988). Gel retardation at low pH resolves trp repressor-DNA complexes for quantitative study. *Proc. natn. Acad. Sci. U.S.A.* **85**, 975–979.

Carey, J. (1989). trp Repressor arms contribute binding energy without occupying unique locations on DNA. *J. biol. Chem.* **264**, 1941–1945.

Carmichael, V. E., Dutton, P. J., Fyles, T. M., James, T. D., Swan, J. A. & Zojaji, M. (1989). Biomimetic ion transport: a functional model of a unimolecular ion channel. *J. Am. Chem. Soc.* **111**, 767–769.

Caspar, D. (1963). Assembly and stability of the tobacco mosaic virus particle. *Adv. Prot. Chem.* **18**, 37–121.

Cash, D. J. & Hess, G. P. (1980). Molecular mechanism of acetylcholine-controlled ion translocation across cell membranes. *Proc. natn. Acad. Sci U.S.A.* **77**, 842–846.

Changeux, J. P. (1961). The feedback mechanism of biosynthetic L-threonine deaminase by L-isoleucine. *Cold Spring Harb. Symp. quant. Biol.* **26**, 313–318.

Changeux, J. P., Gerhardt, J. C. & Schachman, H. K. (1968). Allosteric interaction in aspartate transcarbamylase. 1. Binding of specific ligands to the native enzyme and its isolated subunits. *Biochemistry* **7**, 531–538.

Changeux, J. P., Giraudat, J. & Dennis, M. (1987). The nicotinic acid acetylcholine receptor: molecular architecture of a ligand-regulated ion channel. *Trends pharmac. Sci.* **8**, 459–465.

Changeux, J. P. & Rubin, M. M. (1968). Allosteric interactions in aspartate transcarbamylase. III. Interpretation of experimental data in terms of the model of Monod, Wyman & Changeux. *Biochemistry* **7**, 553–560.

Cheng, X. & Schoenborn (1990). Hydration in protein crystals: a neutron diffraction analysis of carbonmonoxymyoglobin. *Acta Cryst.* (in the Press).

Chothia, C. & Lesk, A. M. (1985). Helix movements in proteins. *Trends in Biochem. Sci.* **10**, 116–118.

Chu, A. H., Turner, B. W. & Ackers, G. K. (1984). Effects of protons on the oxygen-linked subassembly in human hemoglobin. *Biochemistry* **23**, 604–617.

Cohen, G. & Jacob, F. (1959). Sur la répression de la synthèse des enzymes intervenants dans la formation du tryptophane chez *Escherichia coli*. *C. r. hebd. séanc. Acad. Sci., Paris* **248**, 3490–3492.

Connelly, P. R., Gill, S. J., Miller, K. I., Zhou, G. & van Holde, K. E. (1989). Identical linkage and cooperativity of oxygen and carbon monoxide binding to *Octopus dofleini* hemocyanin. *Biochemistry* **28**, 1835–1843.

Crick, F. H. C. & Orgel, L. E. (1964). The theory of interallelic complementation. *J. molec. Biol.* **8**, 161–165.

Crothers, D. M. & Metzger, H. (1972). The influence of polyvalency on the binding properties of antibodies. *Immunochemistry* **9**, 341–357.

Dalvit, C. & Wright, P. E. (1987). Assignment of resonances in the ^{1}H nuclear magnetic resonance spectrum of the carbonmonoxide complex of sperm whale myoglobin by phase-sensitive two-dimensional techniques. *J. molec. Biol.* **194**, 313–327.

Derewenda, Z., Dodson, G., Emsley, P., Harris, D., Nagai, K., Perutz, M. F. & Renaud, J.-P. (1989). The stereochemistry of CO binding to normal human adult and Cowtown hemoglobins. *J. molec. Biol.* (in the Press).

Dickerson, R. E. & Geis, I. (1983). *Haemoglobin: Structure, Function, Evolution, Pathology*. Kenlo Park: Benjamin/Cummings.

Di Gabriele, A. D., Sanderson, M. R. & Steitz, T. A. (1989). Crystal lattice packing is important in determining the bend of a DNA dodecamer containing an adenine tract. *Proc. natn. Acad. Sci. U.S.A.* **86**, 1816–1920.

Dohi, D., Sugita, Y. & Yoneyama, Y. (1973). The self-association and oxygen equilibrium of hemoglobin from the lamprey *Entosphenus japonicus*. *J. biol. Chem.* **248**, 2354–2363.

Eisenberg, D., Almassy, R. J., Janson, C. A. Chapman, M. S., Shuh, S. W., Cascio, D. & Smith, W. W. (1987). Some evolutionary relationships of the primary biological catalysts glutamine synthetase and RuBisCo. *Cold Spring Harb. Symp. quant. Biol.* **52**, 483–490.

Eisenstein, E., Markby, D. W. & Schachman, H. K. (1989). Changes in stability and allosteric properties of aspartate transcarbamyolase resulting from amino acid substitutions in the zinc-binding domain of the regulatory chains. *Proc. natn. Acad. Sci. U.S.A.* **86**, 3094–3098.

Elam, W. T., Stern, E. A., McCallum, J. D. & Sanders-Loehr, J. (1983). An X-ray absorption study of the binuclear iron center in deoxyhemerythrin. *J. Am. Chem. Soc.* **105**, 1919–1923.

Elber, R. & Karplus, M. (1990). Molecular dynamics simulations of myoglobin. Submitted to *J. Amer. Chem. Soc.*

Ellerton, H. D., Ellerton, N. F. & Robinson, H. A. (1983). Hemocyanin – a current perspective. *Progr. Biophys. Molec. Biol.* **41**, 143–248.

Englander, S. W. & Kallenbach, N. R. (1984). Hydrogen exchange and structural dynamics of proteins and nucleic acids. *Q. Rev. Biophys.* **16**, 521–565.

Evans, D. R., Styliani, C. P.-L. & Lipscomb, W. N. (1975). Isolation and properties of a species produced by the partial dissociation of aspartate transcarbamylase from *Escherichia coli*. *J. biol. Chem.* **250** (10), 3571–3583.

Evans, P. R. & Hudson, P. J. (1979). Structure and control of phosphofructokinase from *Bacillus stearothermophilus*. *Nature* **279**, 500–504.

Evans, P. R., Farrants, G. W. & Lawrence, M. C. (1986). Crystallographic structure of allosterically inhibited phosphofructokinase at 7 Å resolution. *J. molec. Biol.* **191**, 713–720.

Fermi, G. & Perutz, M. F. (1981). *Atlas of Molecular Structures in Biology : Haemoglobin & Myoglobin*. Oxford: Clarendon Press.

Fersht, A. R.. Leatherbarrow, R. J. & Wells, T. N. C. (1986). Structure and activity of the tyrosyl-tRNA synthetase: the hydrogen bond in catalysis and specificity. *Phil. Trans. R. Soc. Lond.* A**317**, 305–320.

Fincham, J. R. S. & Day, P. R. (1963). *Fungal Genetics*. Oxford: Blackwell Scientific Publications.

Fletterick, R. J. & Madsen, N. B. (1980). The structures and related functions of phosphorylase a. *A. Rev. Biochem.* **49**, 31–61.

Fletterick, R. J. & Sprang, S. R. (1982). Glycogen phosphorylase structures and function. *Acc. Chem. Res.* **15**, 361–369.

Foote, J. & Schachman, H. K. (1985). Homotropic effects in aspartate transcarbamylase: What happens when the enzyme binds a single molecule of the bisubstrate analogue *N*-phosphonacetyl-L-aspartate? *J. molec. Biol* **186**, 175–184.

Frederick, C. A., Grable, J., Melia, M., Samudzi, C., Jen-Jacobson, L., Wang, B. C., Greene, P., Boyer, H. W. & Rosenberg, J. M. (1984). Kinked DNA in crystalline complex with *Eco* RI endonuclease. *Nature* **309**, 327–331.

Frey, J. G., Eisenberg, D. & Eiserling, F. A. (1975). Glutamine synthetase forms three- and seven-stranded cables. *Proc. natn. Acad. Sci. U.S.A.* **72**, 3402–3406.

Gaykema, W. P. J., Hol, W. G. J., Vereijken, J. M., Soeter, N. M., Bak, H. J. & Beintema, J. J. (1984). 3·2 Å structure of the copper-containing, oxygen-carrying protein *Panulirus interruptus* haemocyanin. *Nature* **309**, 23–29.

Gelin, R. G., Lee, A. W.-M. & Karplus, M. (1983). Hemoglobin tertiary structural change on ligand binding. Its role in the cooperative mechanism. *J. molec. Biol.* **171**, 489–559.

Gerhardt, J. C. & Pardee, A. B. (1962). The enzymology of control by feedback inhibition. *J. biol. Chem.* **237**, 891–896.

Gerhardt, J. C. & Schachman, H. K. (1968). Allosteric interactions in aspartate transcarbamylase II. Evidence for different conformational states of the protein in the presence and absence of specific ligands. *Biochemistry* **7**, 538–552.

Gibbons, I., Ritchey, J. M. & Schachman, H. K. (1976). Concerted allosteric transitions in hybrids of aspartate transcarbamylase containing different arrangements of active and inactive sites. *Biochemistry* **15**, 1324–1330.

Gibbons, I., Yang, Y. R. & Schachman, H. K. (1974). Cooperative interactions in aspartate transcarbamylase. I. Hybrids composed of native and chemically inactivated catalytic chains. *Proc. natn. Acad. Sci. U.S.A.* **71**, 4452–4456.

Gill, S. J., Di Cera, E., Doyle, M. L. & Robert, C. H. (1988). New twists on an old story: hemoglobin. *Trends biochem. Sci.* **13**, 465–467.

Giraudat, J., Dennis, M., Heitmann, T., Haumont, P. T., Lederer, F. & Changeux, J. P. (1987). Structure of the high-affinity binding sites for non-competitive blockers of the acetylcholine receptor: [^{3}H]chlorpromazine labels homologous residues in the β and δ chains. *Biochemistry* **26**, 2410–2418.

Goldsmith, E., Sprang, S. & Fletterick, R. J. (1982). Structure of maltoheptaose by difference Fourier methods and a model for glycogen. *J. molec. Biol.* **156**, 411–423.

Goldsmith, E. J., Sprang, S. R., Hamlin, R., Xuong, N. & Fletterick, R. J. (1989). Domain separation in the activation of glycogen phosphorylase a. *Science* **245**, 528–532.

Gopalakrishnan, P. V. & Karush, F. (1974). Antibody affinity. VII. Multivalent interaction of anti-lactoside antibody. *J. Immunol.* **113**, 769–778.

Gouaux, E. J. & Lipscomb, W. N. (1988). Three-dimensional structure of carbamoylphosphate and succinate bound to aspartate carbamoyltransferase. *Proc. natn. Acad. Sci. U.S.A.* **85**, 4205–4208.

Graves, D. J. & Wang, J. H. (1972). α-glucan phosphorylases – chemical and physical basis of catalysis and regulation. In *The Enzymes*, (ed. P. D. Boyer) vol. 7, pp. 435–482. New York: Academic Press.

Gunsalus, R. P., Gunsalus, M. A. & Gunsalus, G. L. (1986). Intracellular trp repressor levels in Escherichia coli. *J. Bacteriol.* **167**, 272–278.

Hajdu, J., Acharya, K. R., Stuart, D. I., McLaughlin, P. J., Barford, D., Oikonomakos, N. G., Klein, H. & Johnson, L. N. (1987). Catalysis in the crystal: synchrotron radiation studies with glycogen phosphorylase b. *EMBO J.* **6**, 539–545.

Heidmann, T., Beuchardt, J. Neumann, E. & Changeux, J. P. (1983). Rapid kinetics of agonist binding and permeability response analysed in parallel on acetylcholine receptor-rich membranes from *Torpedo marmorata*. *Biochemistry* **22**, 5452–5459.

Heidner, E. J., Frey, T. G. Held, J., Weissman, L. J., Fenna, R. E., Lei, M., Harel, M., Kabsch, H., Sweet, R. M. & Eisenberg, D. (1978). New crystal forms of glutamine synthetase and implications for the molecular structure. *J. molec. Biol.* **122**, 163–173.

HELLINGA, H. W. & EVANS, P. R. (1987). Mutations in the active site of *Escherichia coli* phosphofructokinase. *Nature* **327**, 437–439.

HELMREICH, E. J. M. & KLEIN, H. W. (1980). The role of pyridoxal phosphate in the catalysis of glycogen phosphorylase. *Angew. Chem. Int. Ed. Eng.* **19**, 441–455.

HENDRICKSON, W. A. & LOVE, W. E. (1971). Structure of lamprey haemoglobin. *Nature New Biol.* **232**, 197–203.

HERVÉ, J. (1989). Aspartate transcarbamylase from *E. coli*. In *Allosteric Enzymes* (ed. G. Hervé) CRC Press.

HERVÉ, G., MOODY, M. F., TAVE, P., VACHETTE, P. & JONES, P. T. (1985). Quaternary structure changes in aspartate transcarbamylase by X-ray scattering. *J. molec. Biol.* **185**, 189–199.

HONZATKO, R. B., CRAWFORD, J. L., MONACO, H. L., LADNER, J. E., EDWARDS, B. F. P., EVANS, D. R., WARREN, S. G., WILEY, D. C., LADNER, R. C. & LIPSCOMB, W. N. (1982). Crystal and molecular structure of native and CTP-liganded aspartate carbamoyltransferase from *Escherichia coli*. *J. molec. Biol.* **160**, 219–263.

HONZATKO, R. B. & LIPSCOMB, W. N. (1982). Interactions of phosphate ligands with *Escherichia coli* aspartate carbamoyltransferase in the crystalline state. *J. molec. Biol.* **160**, 265–286.

HOWLETT, G. J., BLACKBURN, M. N., COMPTON, J. G. & SCHACHMAN, H. K. (1977). Allosteric regulation of aspartate transcarbamylase. Analysis of the structural and functional behaviour in terms of a two-state model. *Biochemistry* **16**, 5091–5099.

HOWLETT, G. J. & SCHACHMAN, H. K. (1977). Allosteric regulation of aspartate transcarbamylase. Changes in the sedimentation coefficient promoted by the bisubstrate analogue *N*-(phosphonacetyl)-L-aspartate. *Biochemistry* **16**, 5077–5083.

IMAI, K., (1982). *Allosteric Effects in Haemoglobin*. Cambridge University Press.

JARVEST, R. L., LOWE, G. & POTTER, B. V. L. (1981). The stereochemical course of phosphoryl transfer catalyzed by *Bacillus stearothermophilus* and rabbit-muscle phosphofructokinase with a chiral [^{16}O, ^{17}O, ^{18}O,]phosphate ester. *Biochem.J.* **199**, 427–432.

JOACHIMIAK, A. J., KELLEY, R. L., GUNSALUS, R. P., YANOFSKY, C. & SIGLER, P. B. (1983). Purification and characterization of the trp aporepressor. *Proc. natn. Acad. Sci. U.S.A.* **80**, 668–672.

JOHNSON, B. A., BONAVENTURA, C. & BONAVENTURA, J. (1984). Allosteric modulation of *Callinectes sapidus* haemocyanin by lactate. *Biochemistry* **23**, 872–878.

JOHNSON, K. A., OLSON, J. S. & PHILLIPS, G. N. JR. (1989). The structure of myoglobin-ethyl isocyanide: histidine as a swinging door for ligand entry. *J. molec. Biol.* **207**, 459–463.

JOHNSON, L. N., HAJDU, J., ACHARYA, K. R., STUART, D. I., MCLAUGHLIN, P. J., OIKONOMAKOS, N. G. & BARFORD, D. (1989). Glycogen phosphorylase b. In *Allosteric Proteins* (ed. G. Hervé), CRC Press.

JULLIEN, L. & LEHN, J.-M. (1988). The "Chundle" approach to molecular channels: synthesis of a macrocycle-based molecular bundle. *Tetrahedron Lett.* **29** (31), 3803–3806.

KANTROWITZ, E. R. & LIPSCOMB, W. N. (1988). *Escherichia coli* aspartate transcarbamylase: the relations between structure and function. *Science* **241**, 669–674.

KANTROWITZ, E. R. & LIPSCOMB, W. N. (1990). *Escherichia coli* aspartate transcarbamoylase: Part III: The molecular basis for a concerted allosteric transition. *Trends Biochem. Sci.* (in the Press).

KARUSH, F. (1978). The affinity of antibody: range, variability, and the role of multivalence. In *Immunoglubulins* (ed. G. W. Littman and R. A. Good), pp. 85–115. Plenum.

KE, H.-M., HONZATKO, R. B. & LIPSCOMB, W. N., (1984). Structure of unligated aspartate carbamoyltransferase of *Escherichia coli* at 2·6 Å resolution. *Proc. natn. Acad. Sci. U.S.A.* **81**, 4037–4040.

KE, H., LIPSCOMB, W. N., CHO, Y. & HONZATKO, R. B. (1988). Complex of *N*-phosphonyl-L-aspartate with aspartate carbamoyltransferase. X-ray refinement, analysis of conformational changes and catalytic and allosteric mechanisms. *J. molec. Biol.* **204**, 725–747.

KELLY, R. & YANOFSKY, C. (1985). Mutational studies with the trp repressor of *E. coli* support the helix-turn-helix model of repressor recognition of operator DNA. *Proc. natn. Acad. Sci. U.S.A.* **82**, 483–487.

KILMARTIN, J. V. (1974). Influence of DPG on the Bohr effect of human haemoglobin. *FEBS Lett.* **38**, 147–148.

KILMARTIN, J. V., BREEN, J. J., ROBERTS, G. C. K. & HO, C. (1973). Direct measurement of the p*K* values of an alkaline Bohr group in human haemoglobin. *Proc. natn. Acad. Sci. U.S.A.* **70**, 1246–1249.

KIM, K., FETTINGER, J., SESSLER, J. L., CYR, M., HUGDAHL, J., COLLMAN, J. P. & IBERS, J. A. (1989). Structural characterisation of a sterically encumbered iron(II) porphyrin CO complex. *J. Am. Chem. Soc.* **111**, 403–405.

KLUG, A. & RHODES, D. (1987). 'Zinc fingers': a novel protein motif for nucleic acid recognition. *Trends biochem. Sci.* **12**, 464–469.

KNOWLES, J. (1980). Enzyme-catalyzed phosphoryl transfer reactions. *An. Rev. Biochem.* **49**, 877–918.

KOSHLAND, D. E. (1959). In *The Enzymes*, 2nd ed. vol. 1 (ed. P. D. Boyer, H. Lardy & K. Myrbäck), pp. 305–346. New York: Academic Press.

KOSHLAND, D. E., NEMETHY, G. & FILMER, D. (1966). Comparison of experimental binding data and theoretical models in proteins containing subunits. *Biochemistry* **5**, 365–385.

KOTLARZ, D. & BUC, H. (1982). Phosphofructokinases from *Escherichia coli*. *Methods Enzymol.* **90**, 60–70.

KRAUSE, K. L., VOLZ, K. W. & LIPSCOMB, W. N. (1987). 2·5 Å structure of aspartate carbamoyltransferase complexed with the bisubstrate analog *N*-(phosphonacetyl)-L-aspartate. *J. molec. Biol.* **193**, 527–553.

KREBS, E. G. (1986). The enzymology of control by phosphorylation. In *The Enzymes*, 3rd ed. vol. 17 (ed. P. D. Boyer and E. G. Krebs), pp. 3–20. New York: Academic Press.

KRÜSE, J., KRÜSE, K. M., WITZ, J., CHAUVIN, C., JACROT, B. & TARDIEU, A. (1982). Divalent ion-dependent reversible swelling of tomato bushy stunt virus and organisation of the expanded virion. *J. molec. Biol.* **162**, 393–417.

KUIPER, H., ANTONINI, E. & BRUNORI, M. (1977). Kinetic control of cooperativity in the oxygen binding of *Panulirus interruptus* haemocyanin. *J. molec. Biol.* **116**, 569–576.

KUIPER, H., FORLANI, L., CHIANCONE, E., ANTONINI, E., BRUNORI, M. & WYMAN, J. (1979). Multiple linkage in *Panulirus interruptus*. *Biochemistry* **18**, 5849–5854.

KUIPER, H., GAASTRA, W., BEINTEMA, J. J., VAN BRUGGEN, E. F. J., SCHEPMAN, A. M. H. & DRENTH, J. (1975). Subunit composition, X-ray diffraction, amino acid analysis and oxygen binding behaviour of *Panulirus interruptus* hemocyanin. *J. molec. Biol.* **99**, 619–629.

KURIYAN, J., WILZ, S., KARPLUS, M. & PETSKO, G. A. (1986). X-ray structure and refinement of carbonmonoxy (FeII)-myoglobin at 1·5 Å resolution. *J. molec. Biol.* **192**, 133–154.

LADJIMI, M. M. & KANTROWITZ, E. R. (1988). A possible model for the concerted transition in *Escherichia coli* aspartate transcarbamylase as deduced from site-directed mutagenesis studies. *Biochemistry* **27**, 276–283.

LADJIMI, M. M., MIDDLETON, S. A., KELLERHER, K. S. & KANTROWITZ, E. R. (1988). Relationship between domain closure and binding, catalysis and regulation in *Escherichia coli* aspartate transcarbamylase. *Biochemistry* **27**, 268–276.

LAKOWICZ, J. R. & WEBER, G. (1973). Quenching of protein fluorescence by oxygen. Detection of structural fluctuations in proteins on the nanosecond time scale. *Biochemistry* **12**, 4171–4179.

LALEZARI, I., RAHBAR, S., LALEZARI, P., FERMI. G. & PERUTZ, M. F. (1988). LR16, a compound with potent effects on the oxygen affinity of hemoglobin, on blood cholesterol, and on low density lipoprotein. *Proc. natn. Acad. Sci. U.S.A.* **85**, 6117–6121.

LAMY, J., LECLERC, M., SIZARET, P.-Y., LAMY, J., MILLER, K. I., MCPARLAND, R. & VAN HOLDE, K. E. (1987). *Octopus dofleini* hemocyanin: structure of the seven-domain polypeptide chain. *Biochemistry* **26**, 3509–3518.

LAU, F. T.-K. & FERSHT, A. R. (1987). Conversion of allosteric inhibition to activation in phosphofructokinase by protein engineering. *Nature* **326**, 811–812.

LAU, F. T.-K., FERSHT, A. R., HELLINGA, H. W. & EVANS, P. R. (1987). Site-directed mutagenesis in the effector site of *Escherichia coli* phosphofructokinase. *Biochemistry* **26**, 4143–4148.

LAWSON, C. & SIGLER, P. B. (1988). The structure of *trp* pseudorepressor at 1·65 Å shows why indol propionate acts as a *trp* 'inducer'. *Nature* **333**, 869–871.

LAWSON, C. L., ZHANG, R.-g., SCHEVITZ, R. W., OTWINOWSKI, Z., JOACHIMIAK, A. & SIGLER, P. B. (1988). Flexibility of the DNA-binding domains of *trp* repressor. *Proteins* **3**, 18–31.

LEAR, J. D., WASSERMAN, Z. R. & DEGRADE, W. F. (1988). Synthetic amphiphilic peptide models for protein ion channels. *Science* **240**, 1177–1181.

LEONARD, R. J., LABARCA, C. G., CHARNET, P., DAVIDSON, N. & LESTER, H. A. (1988). Evidence that the M2 spanning region lines the ion channel pore of the nicotinic receptor. *Science* **242**, 1578–1581.

LIDDINGTON, R., DEREWENDA, Z., DODSON, G. & HARRIS, D. (1988). Structure of liganded T-state of haemoglobin identifies the origin of cooperative oxygen binding. *Nature* **331**, 725–728.

LINZEN, B., SOETER, N. M., RIGGS, A. F., SCHNEIDER, H.-J., SCHARTAU, W., MOORE, M. D., YOKOTA, E., BEHRENS, P. Q., NAKASHIMA, H., TAKAGI, T., NEMOTO, T., VEREIJKEN, J. M., BAK, H. J., BEINTEMA, J. J., VOLBEDA, A., GAYKEMA, W. P. J. & HOL. W. G. J. (1985). The structure of arthropod haemocyanins. *Science, N.Y.* **229**, 519–524.

LOUIE, G., TRAN, T., ENGLANDER, J. J. & ENGLANDER, S. W. (1988). Allosteric energy at the hemoglobin β-chain C-terminus studied by hydrogen exchange. *J. molec. Biol.* **201**, 755–764.

MCGEOCH, D., MCGEOCH, J. & MORSE, D. (1973). Synthesis of tryptophan operon RNA in a cell-free system. *Nature New Biol.* **245**, 137–140.

MCLAUGHLIN, P. J., STUART, D. I., KLEIN, H. W., OINOMAKOS, J. G. & JOHNSON, L. N.

(1984). Substrate cofactor interactions for glycogen phosphorylase b: a binding study in the crystal with heptenitol and heptulose-2-phosphate. *Biochemistry* **23**, 5862–5873.

MADSEN, N. B. (1986). Glycogen phosphorylase. In *The Enzymes*, 3rd ed., vol. 17 (ed. P. D. Boyer and E. G. Krebs), pp. 366–394. New York: Academic Press.

MANWELL, C. (1960). Oxygen equilibrium of the brachiopod *Lingula* hemerythrin. Science **132**, 550–551.

MARMORSTEIN, R. Q. & SIGLER, P. B. (1988). Structure and mechanism of the *trp* repressor/operator system. In *Nucleic Acids and Molecular Biology* (ed. F. Eckstein). Heidelberg. Springer-Verlag.

MARMORSTEIN, R. Q., JOACHIMIAK, A., SPRINZL, M. & SIGLER, P. B. (1987). The structural basis for the interaction between L-tryptophan and the *Escherichia coli trp* aporepressor. *J. biol. Chem.* **262**, 4922–4927.

MATTHEWS, A. J., ROHLFS, R. J., OLSON, J. S., TAME, J., RENAUD, J.-P. & NAGAI, K. (1989). The effects of E7 and E11 mutations on the kinetics of ligand binding to the R-state human hemoglobin. *J. Biol. Chem.* **264**, 16573–16583.

MATSUKAWA, S., ITATANI, Y., MAWATARI, K., SHIMOKAWA, Y. & YONEYAMA, Y. (1978). Quantitative evaluation for the role β-146 His and β-143 His residues in the Bohr effect of human haemoglobin in the presence of 0·1 M chloride ion. *J. biol. Chem.* **259**, 11479–11486.

MESSANA, C., CERDONIO, M., SHENKIN, P., NOBLE, R. W., FERMI, G., PERUTZ, R. N. & PERUTZ, M. F. (1978). Influence of quaternary structure of the globin on thermal spin equilibria in different methaemoglobin derivatives. *Biochemistry* **17**, 3652–3662.

MIDDLETON, S. A. & KANTROWITZ, E. R. (1988). Function of arginine-234 and aspartic acid-271 in domain closure, cooperativity, and catalysis in *Escherichia coli* aspartate transcarbamylase. *Biochemistry* 27, 8653–8660.

MILLER, K. I. (1985). Oxygen equilibria of *Octopus dofleini* hemocyanin. *Biochemistry* **24**, 4582–4586.

MOMENTEAU, M., SCHEIDT, W. R., EIGENBROT, C. W. & REED, C. A. (1988). A deoxymyoglobin model with a sterically unhindered axial imidazole. *J. Am. Chem. Soc.* **110**, 1207–1215.

MONOD, J., CHANGEUX, J. P. & JACOB, F. (1963). Allosteric proteins and molecular control systems. *J. molec. Biol.* **6**, 306–329.

MONOD, J. & COHEN-BAZIRE, G. (1953). L'effet d'inhibition spécifique dans la biosynthèse de la tryptophane-desmase chez *Aerobacter aerogenus. C.r. hebd. séanc. Acad. Sci., Paris* **236**, 530–532.

MONOD, J. & JACOB, F. (1961). Genetic regulation mechanisms in the synthesis of proteins. *J. molec. Biol.* **3**, 318–356.

MONOD, J., WYMAN, J. & CHANGEUX, J. P. (1965). On the nature of allosteric transitions: a plausible model. *J. molec. Biol.* **12**, 88–118.

NAMBA, K., PATTANAYEK, R. & STUBBS, G. (1989). Visualization of protein-nucleic acid interactions in a virus: refined structure of intact tobacco mosaic virus at 2·9 Å resolution by X-ray diffraction. *J. molec. Biol.* **208**, 307–325.

NEWELL, J. O., MARKBY, D. W. & SCHACHMAN, H. K. (1989). Cooperative binding of the bisubstrate analog *N*-(phosphonacetyl)-L-aspartate to aspartate transcarbamylase and the heterotropic effects of ATP and CTP. *J. biol. Chem.* **264**, 2476–2481.

OIKI, S., DANHO, W., MADISON, V. & MONTAL, M. (1988). M2 δ, a candidate for the structure lining the ionic channel of the nicotinic cholinergic receptor. *Proc. natn. Acad. Sci. U.S.A.* **85**, 8703–8707.

OLSON, J. S., MATHEWS, A. J., ROHLFS, R. J., SPRINGER, B. A., EDELBERG, K. D., SLIGER, S. G., TAME, J., RENAUD, J.-P. & NAGAI, K. (1988). The role of the distal histidine in myoglobin and haemoglobin. *Nature* **336**, 265–266.

ONAN, K., REBEK, J., COSTELLO, T. & MARSHALL, L. (1983). Allosteric effects: structural and thermodynamic origins and binding cooperativity in a subunit model. *J. Am. Chem. Soc.* **105**, 6759–6760.

OTWINOWSKI, Z., SCHEVITZ, R. W., ZHANG, R.-g., LAWSON, C. L., JOACHIMIAK, A., MARMORSTEIN, R. Q., LUISI, B. & SIGLER, P. B. (1988). The crystal structure of the *trp* repressor/operator complex at atomic resolution. *Nature* **335**, 321–329.

PALM, D., KLEIN, H. W., SCHINZEL, R., BEUHNER, M. & HELMREICH, E. J. M. (1990). The role of pyroxidal 5′-Phosphate in glycogen phosphorylase catalysis. *Perspectives in Biochem.* (in the Press).

PERRELLA, M., SABIONEDDA, L., SAMAJA, M. & ROSSI-BERNARDI, L. (1986). The intermediate compounds between human hemoglobin and carbon monoxide at equilibrium and during approach to equilibrium. *J. biol. Chem.* **261**, 8391–8396.

PERRELLA, M., COLOSIMO, A., BENAZZI, L., SAMAJA, M. & ROSSI-BERNARDI, L. (1988). Intermediate compounds between hemoglobin and carbonmonoxide under equilibrium conditions. Symposion on Oxygen Binding and Heme Proteins, Asilomar Conference Grounds, Pacific Grove, California.

PERUTZ, M. F. & MATHEWS, F. S. (1966). An X-ray study of azide methaemoglobin. *J. molec. Biol.* **21**, 199–202.

PERUTZ, M. F. (1970). Stereochemistry of cooperative effects in haemoglobin. *Nature* **228**, 726–739.

PERUTZ, M. F. (1979). Regulation of oxygen affinity of hemoglobin: influence of structure of the globin on the heme. *A. Rev. Biochem.* **48**, 327–386.

PERUTZ, M. F. (1987). Molecular anatomy, physiology and pathology of hemoglobin. In *The Molecular Basis of Blood Diseases* (ed. G. Stamatoyannopoulos, A. W. Nienhuis, P. Leder and P. W. Majerus), pp. 127–178. Philadelphia: W. B. Saunders.

PERUTUZ, M. F. (1988). Allosteric enzymes: control by phosphorylation. *Nature* **336**, 202–203.

PERUTZ, M. F., FERMI, G., ABRAHAM, D. J., POYART, C. & BURSAUX, E. (1986). Hemoglobin as a receptor of drugs and peptides: X-ray studies of the stereochemistry of binding. *J. Am. Chem. Soc.* **108**, 1064–1078.

PERUTZ, M. F., FERMI, G., LUISI, B., SHAANAN, B. & LIDDINGTON, R. C. (1987). Stereochemistry of cooperative effects in hemoglobin. *Accs. Chem. Res.* **20**, 309–321.

PERUTZ, M. F., KILMARTIN, J. V., NISHIKURA, K., FOGG, J. H., BUTLER, P. J. G. & ROLLEMA, H. S. (1980). Identification of residues contributing to the Bohr effect of human haemoglobin. *J. molec. Biol.* **138**, 649–670.

PERUTZ, M. F., SANDERS, J. K. M., CHENERY, D. H., NOBLE, R. W., PENNELLY, R. R. FUNG, L. W.-M., HO, C., GIANNINI, I., PÖRSCHKE, D. & WINCKLER, H. (1978). Interactions between the quaternary structure of the globin and the spin state of the heme in ferric mixed spin derivatives of hemoglobin. *Biochemistry* **17**, 3640–3652.

PHILLIPS, S. E. V. & SCHOENBORN, B. P. (1981). Neutron diffraction reveals oxygen-histidine hydrogen bond in myoglobin. *Nature* **292**, 81–82.

PHILLIPS, S. E. V., MANSFIELD, I., PARSONS, I., DAVIDSON, B. E., RAFFERTY, J. B., SOMERS, W. S., MARGARITA, D., COHEN, G. N., SAINT-GIRONS, D. & STOCKLEY, P. G. (1989). Cooperative tandem binding of *met* repressor of *E. coli*. *Nature* **341**, 711–715.

PHILO, S. & DREYER, U. (1985). Quarternary structure has little influence on spin states in mixed-spin human methemoglobins. *Biochemistry* **24**, 2985–2991.

Ptashne, M. (1986). *A Genetic Switch.* Cambridge, Mass.: Cell Press; Oxford: Blackwell Scientific Publications.

Rafferty, J. B., Somers, W. S., St.-Girons, I. & Phillips, S. E. V. (1989). Three-dimensional crystal structures of *E. coli* met repressor with and without co-repressors. *Nature* **341**, 705–710.

Rebek, J., Wattley, R. V., Costello, T., Gadwood, R. & Marshall, L. (1981). Allosterische Effekte: Bindungskooperativität in einer Modellverbindung mit Untereinheiten. *Angew. Chemie* **93**, 584–585.

Reem, R. C. & Solomon, E. I. (1987). Spectroscopic studies of the binuclear ferrous active site of deoxyhemerythrin; coordination number and probable bridging ligands for the native and ligand-bound forms. *J. Am. Chem. Soc.* **109**, 1216–1226.

Remington, S., Wiegand, G. & Huber, R. (1982). Crystallographic refinement and atomic models of two different forms of citrate synthase at 2·7 Å and 1·7 Å resolution. *J. molec. Biol.* **158**, 111–152.

Richardson, D. E., Reem, R. C. & Solomon, E. I. (1983). Cooperativity in oxygen binding to *Lingula reevii* hemerythrin: spectroscopic comparisons to the sipunculid hemerythrin binuclear active site. *J. Am. Chem. Soc.* **105**, 7780–7781.

Ringe, D., Petsko, G. E., Kerr, D. E. & de Montellano, F. R. O. (1984). Reaction of myoglobin with phenylhydrazine – a molecular doorstop. *Biochemistry* **23**, 2–4.

Robert, C. H., Decker, H., Richey, B., Gill, S. J. & Wyman, J. (1987). Nesting: hierarchies of allosteric interactions. *Proc. natn. Acad. Sci. U.S.A.* **84**, 1891–1895.

Robinson, I. K. & Harrison, S. C. (1982). Structure of the expanded state of tobacco bushy stunt virus. *Nature* **297**, 563–568.

Rose, J. K., Squires, C. L., Yanofsky, C., Yang, H.-L. & Zubay, G. (1973). Regulation of *in vitro* transcription of the tryptophan operon by purified RNA polymerase in the presence of partially purified repressor and tryptophan. *Nature New Biol.* **245**, 133–137.

Schachman, H. K. (1988). Can a simple model account for the allosteric transition of aspartate transcarbamylase? *J. biol. Chem.* **263**, 18583–18586.

Scheidt, R. W. & Reed, C. A. (1981). Spin-state/stereochemical relationships in iron porphyrins: implications for the heme proteins. *Chem. Rev.* 81, 543–555.

Schevitz, R. W., Otwinowski, Z., Joachimiak, A., Lawson, C. L. & Sigler, P. B. (1985). The three-dimensional structure of *trp* repressor. *Nature* **317**, 782–786.

Schirmer, T. & Evans, P. R. (1990). The structural basis of the allosteric behaviour of phosphofructokinase. *Nature* (in the Press).

Schleif, R. (1988). DNA binding to proteins. *Science* **241**, 1182–1187.

Shaanan, B. (1983). Structure of human oxyhaemoglobin at 2·1 Å resolution. *J. molec. Biol.* **171**, 31–50.

Shepherdson, M. & Pardee, A. B. (1960). Production and crystallization of aspartate transcarbamylase. *J. biol. Chem.* **235**, 3233–3237.

Sheriff, S., Hendrickson, W. A. & Smith, J. L. (1987). Structure of myohemerythrin in the azidomet state at 1·7/1·3 Å resolution. *J. molec. Biol.* **197**, 273–296.

Shiemke, A. K., Loehr, T. M. & Sanders-Loehr, J. (1984). Resonance Raman study of the μ-oxo bridged binuclear iron center in oxyhemerythrin. *J. Am. Chem. Soc.* **106**, 4951–4956.

Shirakihara, Y. & Evans, P. R. (1988). Crystal structure of the complex of phosphofructokinase from *Escherichia coli* with its reaction products. *J. molec. Biol.* **204**, 973–994.

Somers, W. & Phillips, S. E. V. (1989). Private communication.

Sprang, S. R., Acharya, K. R., Goldsmith, E. J., Stuart, D. I., Varvill, K., Fletterick, R. J., Madsen, N. B. & Johnson, L. N. (1988). Structural changes in glycogen phosphorylase induced by phosphorylation. *Nature* **336**, 215–221.

Springer, B. A., Egeberg, K. D., Sligar, S. G., Rohlfs, R. J., Mathews, A. J. & Olson, J. S. (1989). Discrimination between oxygen and carbon monoxide and inhibition of autoxidation by myoglobin: site-directed mutagenesis of the distal histidine. *J. biol. Chem.* **264**, 3057–3060.

Stadtman, E. R. & Ginsburg, A. (1974). The glutamine synthetase of *Escherichia coli*: structure and control. In *The Enzymes*, vol. 10, pp. 755–808. New York: Academic Press.

Stencamp, R. E., Sieker, L. C., Jensen, L. H., McCallum, J. D. & Sanders-Loehr, J. (1985). Active site structures of deoxyhemerythrin and oxyhemerythrin. *Proc. natn. Acad. Sci. U.S.A.* **82**, 713–716.

Szabo, A. (1978). The kinetics of haemoglobin and transition state theory. *Proc. natn. Acad. Sci. U.S.A.* **75**, 2108–2111.

Tauc, P., Vachette, P., Middleton, S. A. & Kantrowitz, E. R. (1989). Structural consequences of the replacement of Glu-239 by Gln in the catalytic chain of *Escherichia coli* aspartate transcarbamylase. Submitted to *J. molec. Biol.*

Toyoshima, C. & Unwin, N. (1988). Ion channel of acetylcholine receptor reconstructed from images of postsynaptic membranes. *Nature* **336**, 247–250.

Trautmann, A. (1984). A comparative study of the activation of the cholinergic receptor by various agonists. *Proc. R. Soc. Lond.* **218**, 241–251.

Umbarger, E. & Brown, B. (1958). Isoleucine and valine metabolism in *Escherichia coli*. VII. The negative feedback mechanism controlling isoleucine synthesis. *J. biol. Chem.* **233**, 415–420.

Unwin, P. N. T. (1987). Design and action of cell communication channels. *Chemica Scripta* **27** B, 47–51.

Unwin, P. N. T. & Ennis, P. D. (1984). Two configurations of a channel-forming membrane protein. *Nature* **307**, 609–613.

Unwin, P. N. T., Toyoshima, C. & Kubalek, E. (1988). Arrangement of the acetylcholine receptor subunits in the resting and desensitized states, determined by cryoelectron microscopy of crystallized *Torpedo* postsynaptic membranes. *J. Cell Biol.* **107**, 1123–1138.

Unwin, P. N. T. & Zampighi, G. (1980). Structure of the junction between communicating cells. *Nature* **283**, 545.

van Holde, K. E. & Miller, K. I. (1982). Haemocyanins. *Q. Rev. Biophys.* **15**, 1–129.

Volbeda, A. & Hol, W. (1989*a*). Pseudo-twofold symmetry in the copper-binding domain of arthropodan hemocyanins: possible implications for the evolution of oxygen transport proteins. *J. molec. Biol.* **206**, 531–546.

Volbeda, A. & Hol, W. (1989*b*). Crystal structure of hemocyanin from *Panulirus interruptus* refined at 3·2 Å resolution. *J. molec. Biol.* **209**, 249–279.

Watson, H. C. & Kendrew, J. C. (1961). Comparison between the amino-acid sequences of sperm whale myoglobin and of human haemoglobin. *Nature* **190**, 670–672.

Weber, K. (1968). A new structural model of *Escherichia coli* aspartate transcarbamylase and the amino acid sequence of the regulatory chain. *Nature* **218**, 1116–1119.

Wente, S. R. & Schachman, H. K. (1987). Shared active sites in oligomeric enzymes: model studies with defective mutants of aspartate transcarbamylase produced by site-directed mutagenesis. *Proc. natn. Acad. Sci. U.S.A.* **84**, 31–35.

WERNER, W. E., CANN, J. R. & SCHACHMAN, H. K. (1989). Boundary spreading in sedimentation velocity experiments on partially liganded aspartate transcarbamylase. A ligand-mediated isomerization. *J. molec. Biol.* **206**, 231–238.

WERNER, W. E. & SCHACHMAN, H. K. (1989). Analysis of the ligand-promoted conformational change in aspartate transcarbamylase: evidence for a two-state transition from boundary spreading in sedimentation velocity experiments. *J. molec. Biol.* **206**, 221–230.

WILEY, D. C. & LIPSCOMB, W. (1968). Crystallographic determination of the symmetry of aspartate transcarbamylase. *Nature* **218**, 1119–1121.

WYMAN, J. (1967). Allosteric linkage. *J. Am. Chem. Soc.* **89**, 2202–2218.

WYMAN, J. (1984). Linkage graphs: a study in the thermodynamics of macromolecules. *Q. Rev. Biophys.* **17**, 453–488.

YAMASHITA, M. M., ALMASSY, R. J., JANSON, C. A., CASCIO, D. & EISENBERG, D. (1989). Refined atomic model of glutamine synthetase at 3·5 Å resolution. *J. biol. Chem.* **264**, 17681–17690.

YATES, R. A. & PARDEE, A. B. (1956). Control of pyrimidine biosynthesis in *Escherichia coli* by a feedback mechanism. *J. biol. Chem.* **221**, 757–770.

ZHANG, K., STERN, E. A., ELLIS, F., SANDERS-LOEHR, J. & SHIEMKE, A. K. (1988). The active site of hemerythrin as determined by X-ray Absorption Fine Structure. *Biochemistry* **27**, 7470–7479.

ZHANG, R.-g, JOACHIMIAK, A., LAWSON, C. L., SCHEVITZ, R. W., OTWINOWSKI, Z. & SIGLER, P. B. (1987). The crystal structure of *trp* aporepressor at 1·8 Å shows how binding tryptophan enhances DNA affinity. *Nature* **327**, 591–597.

ZUBAY, G., MORSE, D. E., SCHRENK, W. J. & MILLER, J. H. M. (1972). Detection and isolation of the repressor protein for the tryptophan operator of *Escherichia coli*. *Proc. natn. Acad. Sci. U.S.A.* **69**, 1100–1103.

Index

Numbers in italic type indicate pages on which illustrations appear.